C000260469

Food Chains and Food Webs in Aquatic Ecosystems

Food Chains and Food Webs in Aquatic Ecosystems

Editors

Young-Seuk Park
Ihn-Sil Kwak

MDPI • Basel • Beijing • Wuhan • Barcelona • Belgrade • Manchester • Tokyo • Cluj • Tianjin

Editors

Young-Seuk Park
Ecology and Ecological Informatics,
Department of Biology,
Kyung Hee University
Korea

Ihn-Sil Kwak
Department of Ocean Integrated Science,
Chonnam National University
Korea

Editorial Office
MDPI
St. Alban-Anlage 66 4052 Basel,
Switzerland

This is a reprint of articles from the Special Issue published online in the open access journal *Applied Sciences* (ISSN 2076-3417) (available at: https://www.mdpi.com/journal/applsci/special_issues/Food_Chains_Webs_Aquatic_Ecosystems).

For citation purposes, cite each article independently as indicated on the article page online and as indicated below:

LastName, A.A.; LastName, B.B.; LastName, C.C. Article Title. *Journal Name* **Year**, *Volume Number*, Page Range.

ISBN 978-3-0365-0050-8 (Hbk)
ISBN 978-3-0365-0051-5 (PDF)

Contents

About the Editors

Young-Seuk Park is Professor at the Department of Biology, Kyung University, Seoul, Korea. He completed his PhD at Pusan National University. His laboratory studies the effects of environmental changes on biological systems at different hierarchical levels, from molecules through to individuals, populations, and communities using ecological modeling and informatics approaches. In particular, his research is focused on the effects of global changes and alien species on ecosystems, and ecological monitoring and assessment for sustainable ecosystem management. He is interested in the application of computational approaches such as machine learning techniques and advanced statistical methods. He served as President of the Korean Society for Mathematical Biology. He is Associate Editor of numerous scientific journals including *Ecological Modelling, Annales de Limnologie—International Journal of Limnology*, and *Forests*. He has featured as Guest Editor for several international scientific journals, including *Ecological Modelling, Ecological Informatics, Annales de Limnologie—International Journal of Limnology, Inland Waters, Water*, and *Applied Sciences*.

Ihn-Sil Kwak is Professor at the Department of Ocean Integrated Science, Chonnam National University, Yeosu, Korea. Since completing her PhD at Pusan National University, her research has mainly focused on benthos and macroinvertebrate ecology in water and ecotoxicological and gene responses using chironomids. Her scientific experience includes appointments at RIKEN Brain Science (Japan) and Hanyang University (Seoul). Currently, she is Director of the FCF (Food Chain Flow) project team supported by NRF (University Research Center of the Ministry of Education in Korea).

Editorial

Food Chains and Food Webs in Aquatic Ecosystems

Ihn-Sil Kwak [1] and Young-Seuk Park [2,*]

[1] Department of Ocean Integrated Science, Chonnam National University, Yosu 59626, Korea; iskwak@chonnam.ac.kr
[2] Department of Biology, Kyung Hee University, Seoul 02447, Korea
* Correspondence: parkys@khu.ac.kr

Received: 10 July 2020; Accepted: 17 July 2020; Published: 21 July 2020

Abstract: Food chains and food webs describe the structure of communities and their energy flows, and they present interactions between species. Recently, diverse methods have been developed for both experimental studies and theoretical/computational studies on food webs as well as species interactions. They are effectively used for various applications, including the monitoring and assessment of ecosystems. This Special Issue includes six empirical studies on food chains and food webs as well as effects of environmental factors on organisms in aquatic ecosystems. They confirmed the usefulness of their methods including isotope, DNA-barcoding with gut contents, and environmental DNA for biological monitoring and ecosystem assessment.

Keywords: food web; food chain; aquatic ecosystems; monitoring; assessment; environmental DNA; isotope; NGS

---

1. Introduction

It is important to understand the role and function between organisms' interactions in the food web of the aquatic ecosystem. The key biological interaction in the aquatic food web is matter cycling mediated by the food chain, and predation often works as a regulating factor for energy pathways, as well as determining species composition in the ecosystem [1]. In particular, the food sources at the species levels are critical components linking organisms with larger predatory species such as crustaceans and fish within the grazing food chain: rotifers-copepods, micro/macroinvertebrates, and larval/mature fish [2,3]. Consequently, they function as a channel for the flux of organic matter within diverse organism assemblages organized in an intermediate position between the two different food webs, and a way of transferring nutrients and energy from the prey species–predator species loop to higher trophic levels. Thus, the biological prey–predation interactions in the food web are receiving great attention to understand not only the interrelated biological relationships but also the structure and function of aquatic food webs [4].

In recent years, genomic and next-generation sequencing (NGS) technologies have developed rapidly and been applied to the ecological domain. Meta-barcoding techniques have accreted the reliability of identifying specific taxonomic groups of organisms at both species and genus level [5], and environmental DNA (eDNA) have enabled the detection of invisible species in various situations [6,7]. The eDNA approaches have also been used to clarify and understand systematic ecology, particularly biological trophic interaction in both aquatic habitat environments and food webs by collecting information from food sources found in gut contents of species and the excrement of lived organisms. This helps to overcome unidentified limitations of food source analyses, which were based on microscopic analysis [8–11]. At present, it is necessary to develop a method to separate pure gut content from target organisms for a wide range of applications of DNA technology in food source identification. In addition, the most fundamental methodology is to produce a framed "blocking primer", which removes the DNA of the target species from the target gut contents.

On the other hand, changes in temperature, salinity, and metal contamination could affect the uptake, elimination, and biotransformation rates of common organisms [12]. Increasing water temperatures can act as a stressor that impacts the immune and physical responses of aquatic organisms, especially the cascading food chain network linking of plankton–invertebrates–fish communities. Accordingly, a temperature change can significantly affect food chains' related development and the health of aquatic prey and predation organisms. Further, temperature is known to have a significant effect on oxidative stress biomarkers for aquatic organisms. Due to the fact that climate change is expected to result in more frequent and intense heat shock events, it is pertinent to investigate the effect of increasing temperatures on the oxidative stress response of common aquatic organisms.

Oxidative stress is induced by a wide range of environmental components including temperature changes, UV stress, chemical action and oxygen shortages, and an over-production of reactive oxygen species (ROS) in relation to defense mechanisms [13]. The overproduction of ROS can generate oxidative stress which leads to permanent cell damage. Thus, the intracellular accumulation of ROS would not only disrupt the functions of specific tissues and organs but also lead to the premature death of the entire organism [14]. Oxidative stress biomarkers have been widely used in the development of ecological indices and in the assessment of the exposure of aquatic organisms to contaminants from agricultural, industrial, and urban pollution [15]. Oxidative stress is also involved in many biological and pathological processes and normal physiological development [13]. Currently, the study of many molecular markers has been developed in order to understand the physiological response of organisms. Superoxide dismutases (SODs) and catalase (CAT) are important antioxidant enzymes to protect the cell from oxidative damage by ROS. Especially, heat shock protein 90 (HSP90), a highly conserved protein, is a dimer that binds to several cellular proteins, including steroid receptors and protein kinases [16,17]. In aquatic animals, the induction of HSP90 genes and HSPs family has been widely reported in response to cellular stress, including temperature elevation, osmotic stress, hormone stimulation, herbicide toxicity, and viral infections [18,19].

This Special Issue ("Food Chains and Food Webs in Aquatic Ecosystems") aims to share recent information on the study for food chains and food webs in aquatic ecosystems focusing on biological monitoring and assessment of aquatic ecosystems.

2. Papers in This Special Issue

The six papers included in this Issue focus on food chains and food webs in aquatic ecosystems as well as on effects of environmental factors.

To test a hypothesis that differences in invertebrate and fish assemblages in lakes characterized by different trophic conditions determine patterns of variation in the trophic niche width of the fish species depending on their specific feeding habits, Caputi et al. [20] studied the feeding behavior of two omnivorous species (*Anguilla anguilla* and the seabream *Diplodus annularis*), which are ecologically and economically important, using the stable isotope analysis of carbon ($\delta^{13}C$) and nitrogen ($\delta^{15}N$). They found that *A. anguilla* was a generalist in the eutrophic lake, whereas *D. annularis* became more specialist, suggesting that changes in macroinvertebrate and fish community composition affect the trophic strategies of high-trophic level consumers.

Identification of gut contents is helpful to analyze the food source of animals. However, it has several limitations such as small size and fragmentation of gut materials. To overcome these limitations, recently, genomic approaches have been applied to understand the biological interaction including food webs [9,10,21]. Oh et al. [21] proposed a pretreatment method for DNA-barcoding to analyze gut contents of rotifers to provide a better understanding of rotifer food sources and showed that the proposed method is useful to identify food sources of small organisms.

Jo et al. [22] and Kim et al. [23] presented the application of eDNA in costal aquatic ecosystems. Jo et al. [22] determined aquatic community taxonomic composition using eDNA based on an NGS and analyzed the community spatial distribution with regard to environmental parameters and the habitat types. Meanwhile, Kim et al. [23] compared water sampling between the eDNA method and conventional microscopic identification for plankton community composition related to ecological monitoring and assessment of aquatic ecosystems. They found that the eDNA approach provides a wider variety of species composition, while conventional microscopic identification depicts more distinct plankton communities in sites, suggesting that the eDNA approach is a valuable alternative for biological monitoring and diversity assessments in aquatic ecosystems.

Kim et al. [24] assessed the spatial distribution of benthic macroinvertebrate communities responding to their environment such as land use and water quality, and concluded that information such as land use which is easily available characterized effectively the distribution of benthic macroinvertebrates.

To evaluate the toxic effects of di-2-ethylhexyl phthalate (DEHP) on cellular protection in *Macrophthalmus japonicus* crabs, Park et al. [25] identified two stress-related genes and investigated the genomic structure, phylogenetic relationships with other homologous heat shock proteins (HSPs), and transcriptional responses of HSPs under DEHP stress. Their results suggested that DEHP toxicity could disrupt cellular immune protection through transcriptional changes to HSPs in the test organisms.

3. Conclusions

Food chains and food webs describe the structure of communities and their energy flows, and they present interactions between species. Recently, diverse methods have been developed for both experimental studies and theoretical/computational studies. They improve our fundamental ecological knowledge and are effectively used for various applications, including the monitoring and assessment of ecosystems. In particular, ecological monitoring and assessment have advanced in recent decades. Along with the progress of molecular and eDNA techniques, the process of monitoring and assessment has become rapid and accurate. A wide variety of ecological disturbances associated with temperature and salinity changes and other environmental factors are being recognized as threats to the food chain functions of freshwater and marine ecosystems.

This Special Issue included empirical studies on food chains and food webs in aquatic ecosystems. They confirmed the usefulness of their methods including isotope, DNA-barcoding with gut contents, and eDNA for biological monitoring and ecosystem assessment. In further studies, however, theoretical and computational approaches including food web modelling and network analyses are expected to characterize quantitatively the interactions among species as well as ecosystem structures and dynamics through the collaborative works between experimental and computational scientists.

Author Contributions: Conceptualization, I.-S.K. and Y.-S.P.; writing—original draft preparation, I.-S.K. and Y.-S.P.; writing—review and editing, I.-S.K. and Y.-S.P. All authors have read and agreed to the published version of the manuscript.

Funding: This research was funded by the National Research Foundation of Korea (NRF) funded by the Korean government (MSIP) (grant numbers NRF-2019R1A2C1087099 and NRF-2020R1A2C1013936).

Acknowledgments: We would like to thank all contributors in this Special Issue and all reviewers who provided very constructive and helpful comments to evaluate and improve the manuscripts.

Conflicts of Interest: The authors declare no conflict of interest. The funders had no role in the design of the study; in the collection, analyses, or interpretation of data; in the writing of the manuscript, or in the decision to publish the results.

References

1. Carrillo, P.; Medina-Sánchez, J.; Villar-Argaiz, M.; Delgado-Molina, J.; Bullejos Carrillo, F.J. Complex interactions in microbial food webs: Stoichiometric and functional approaches. *Limnetica* **2006**, *25*, 189–204.
2. Wallace, R.L.; Snell, T.W.; Claudia, R.; Thomas, N. *Rotifera Vol. 1: Biology, Ecology and Systematics*, 2nd ed.; Backhuys: Leiden, The Netherlands, 2006.

3. Pree, B.; Larsen, A.; Egge, J.K.; Simonelli, P.; Madhusoodhanan, R.; Tsagaraki, T.M.; Våge, S.; Erga, S.R.; Bratbak, G.; Thingstad, T.F. Dampened copepod-mediated trophic cascades in a microzooplankton-dominated microbial food web: A mesocosm study. *Limnol. Oceanogr.* **2017**, *62*, 1031–1044. [CrossRef]
4. Oh, H.-J.; Jeong, H.-G.; Nam, G.-S.; Oda, Y.; Dai, W.; Lee, E.-H.; Kong, D.; Hwang, S.-J.; Chang, K.-H. Comparison of taxon-based and trophi-based response patterns of rotifer community to water quality: Applicability of the rotifer functional group as an indicator of water quality. *Anim. Cells Syst.* **2017**, *21*, 133–140. [CrossRef]
5. Hebert, P.D.N.; Cywinska, A.; Ball, S.L.; Dewaard, J.R. Biological identifications through DNA barcodes. *Proc. R. Soc. Lond. Ser. B Biol. Sci.* **2003**, *270*, 313–321. [CrossRef]
6. Barnes, M.A.; Turner, C.R. The ecology of environmental DNA and implications for conservation genetics. *Conserv. Genet.* **2016**, *17*, 1–17. [CrossRef]
7. Kress, W.J.; García-Robledo, C.; Uriarte, M.; Erickson, D.L. DNA barcodes for ecology, evolution, and conservation. *Trends Ecol. Evol.* **2015**, *30*, 25–35. [CrossRef] [PubMed]
8. Symondson, W.O.C. Molecular identification of prey in predator diets. *Mol. Ecol.* **2002**, *11*, 627–641. [CrossRef] [PubMed]
9. Carreon-Martinez, L.; Johnson, T.B.; Ludsin, S.A.; Heath, D.D. Utilization of stomach content DNA to determine diet diversity in piscivorous fishes. *J. Fish Biol.* **2011**, *78*, 1170–1182. [CrossRef] [PubMed]
10. Jo, H.; Gim, J.-A.; Jeong, K.-S.; Kim, H.-S.; Joo, G.-J. Application of DNA barcoding for identification of freshwater carnivorous fish diets: Is number of prey items dependent on size class for Micropterus salmoides? *Ecol. Evol.* **2014**, *4*, 219–229. [CrossRef]
11. Jo, H.; Ventura, M.; Vidal, N.; Gim, J.-S.; Buchaca, T.; Barmuta, L.A.; Jeppesen, E.; Joo, G.-J. Discovering hidden biodiversity: The use of complementary monitoring of fish diet based on DNA barcoding in freshwater ecosystems. *Ecol. Evol.* **2016**, *6*, 219–232. [CrossRef]
12. Lydy, M.J.; Belden, J.B.; Ternes, M.A. Effects of temperature on the toxicity of m-parathion, chlorpyrifos, and pentachlorobenzene to *Chironomus tentans*. *Arch. Environ. Contam. Toxicol.* **1999**, *37*, 542–547. [CrossRef]
13. Wu, C.; Zhang, W.; Mai, K.; Xu, W.; Zhong, X. Effects of dietary zinc on gene expression of antioxidant enzymes and heat shock proteins in hepatopancreas of abalone Haliotis discus hannai. *Comp. Biochem. Physiol. Part C Toxicol. Pharmacol.* **2011**, *154*, 1–6. [CrossRef]
14. Carnevali, S.; Petruzzelli, S.; Longoni, B.; Vanacore, R.; Barale, R.; Cipollini, M.; Scatena, F.; Paggiaro, P.; Celi, A.; Giuntini, C. Cigarette smoke extract induces oxidative stress and apoptosis in human lung fibroblasts. *Am. J. Physiol. Lung Cell. Mol. Physiol.* **2003**, *284*, L955–L963. [CrossRef]
15. Vinagre, C.; Madeira, D.; Mendonça, V.; Dias, M.; Roma, J.; Diniz, M.S. Effect of increasing temperature in the differential activity of oxidative stress biomarkers in various tissues of the Rock goby, *Gobius paganellus*. *Mar. Environ. Res.* **2014**, *97*, 10–14. [CrossRef] [PubMed]
16. Pratt, W.B. The role of heat shock proteins in regulating the function, folding, and trafficking of the glucocorticoid receptor. *J. Biol. Chem.* **1993**, *268*, 21455–21458. [PubMed]
17. Csermely, P.; Schnaider, T.; Soti, C.; Prohászka, Z.; Nardai, G. The 90-kDa molecular chaperone family: Structure, function, and clinical applications. A comprehensive review. *Pharmacol. Ther.* **1998**, *79*, 129–168. [CrossRef]
18. Chang, E.S.; Chang, S.A.; Kwllwe, R.; Reddy, P.S.; Snyder, M.J.; Spees, J.L. Quantification of stress in lobsters: Crustacean hyperglycemic hormone, stress proteins, and gene expression. *Am. Zool.* **1999**, *39*, 487–495. [CrossRef]
19. Park, K.; Park, J.; Kim, J.; Kwak, I.-S. Biological and molecular responses of *Chironomus riparius* (Diptera, Chironomidae) to herbicide 2,4-D (2,4-dichlorophenoxyacetic acid). *Comp. Biochem. Physiol. Part C Toxicol. Pharmacol.* **2010**, *151*, 439–446. [CrossRef]
20. Caputi, S.S.; Careddu, G.; Calizza, E.; Fiorentino, F.; Maccapan, D.; Rossi, L.; Costantini, M.L. Changing isotopic food webs of two economically important fish in Mediterranean coastal lakes with different trophic status. *Appl. Sci.* **2020**, *10*, 2756. [CrossRef]
21. Oh, H.-J.; Krogh, P.; Jeong, H.-G.; Joo, G.-J.; Kwak, I.-S.; Hwang, S.-J.; Gim, J.-S.; Chang, K.-H.; Jo, H. Pretreatment method for DNA barcoding to analyze gut contents of rotifers. *Appl. Sci.* **2020**, *10*, 1064. [CrossRef]
22. Jo, H.; Kim, D.-K.; Park, K.; Kwak, I.-S. Discrimination of spatial distribution of aquatic organisms in a coastal ecosystem using eDNA. *Appl. Sci.* **2019**, *9*, 3450. [CrossRef]

23. Kim, D.-K.; Park, K.; Jo, H.; Kwak, I.-S. Comparison of water sampling between environmental DNA metabarcoding and conventional microscopic identification: A case study in Gwangyang Bay, South Korea. *Appl. Sci.* **2019**, *9*, 3272. [CrossRef]
24. Kim, D.-K.; Jo, H.; Park, K.; Kwak, I.-S. Assessing spatial distribution of benthic macroinvertebrate communities associated with surrounding land cover and water quality. *Appl. Sci.* **2019**, *9*, 5162. [CrossRef]
25. Park, K.; Kim, W.-S.; Kwak, I.-S. Effects of di-(2-ethylhexyl) phthalate on transcriptional expression of cellular protection-related HSP60 and HSP67B2 genes in the mud crab *Macrophthalmus japonicus*. *Appl. Sci.* **2020**, *10*, 2766. [CrossRef]

Article

Changing Isotopic Food Webs of Two Economically Important Fish in Mediterranean Coastal Lakes with Different Trophic Status

Simona Sporta Caputi [1], Giulio Careddu [1], Edoardo Calizza [1,2,*], Federico Fiorentino [1], Deborah Maccapan [1], Loreto Rossi [1,2] and Maria Letizia Costantini [1,2]

[1] Department of Environmental Biology, Sapienza University of Rome, via dei Sardi 70, 00185 Rome, Italy; simona.sportacaputi@uniroma1.it (S.S.C.); giulio.careddu@uniroma1.it (G.C.); federico.fiorentino@uniroma1.it (F.F.); deborah.maccapan@uniroma1.it (D.M.); loreto.rossi@uniroma1.it (L.R.); marialetizia.costantini@uniroma1.it (M.L.C.)

[2] CoNISMa, piazzale Flaminio 9, 00196 Rome, Italy

[*] Correspondence: edoardo.calizza@uniroma1.it

Received: 22 February 2020; Accepted: 13 April 2020; Published: 16 April 2020

Abstract: Transitional waters are highly productive ecosystems, providing essential goods and services to the biosphere and human population. Human influence in coastal areas exposes these ecosystems to continuous internal and external disturbance. Nitrogen-loads can affect the composition of the resident community and the trophic relationships between and within species, including fish. Based on carbon ($\delta^{13}C$) and nitrogen ($\delta^{15}N$) stable isotope analyses of individuals, we explored the feeding behaviour of two ecologically and economically important omnivorous fish, the eel *Anguilla anguilla* and the seabream *Diplodus annularis*, in three neighbouring lakes characterised by different trophic conditions. We found that *A. anguilla* showed greater generalism in the eutrophic lake due to the increased contribution of basal resources and invertebrates to its diet. By contrast, the diet of *D. annularis*, which was mainly based on invertebrate species, became more specialised, focusing especially on polychaetes. Our results suggest that changes in macroinvertebrate and fish community composition, coupled with anthropogenic pressure, affect the trophic strategies of high trophic level consumers such as *A. anguilla* and *D. annularis*. Detailed food web descriptions based on the feeding choices of isotopic trophospecies (here Isotopic Trophic Units, ITUs) enable identification of the prey taxa crucial for the persistence of omnivorous fish stocks, thus providing useful information for their management and habitat conservation.

Keywords: food webs; Mediterranean coastal lakes; nitrogen pollution; stable isotopes; trophic relationships; *Anguilla anguilla*; *Diplodus annularis*

1. Introduction

Transitional waters are extremely complex ecosystems [1–3]. The Water Framework Directive of the European Communities (European Communities, 2000. Directive 2000/60/EC of the European Parliament and of the Council of 23 October 2000) defines them as "superficial bodies of water near the mouths of rivers which have a partially saline character due to their proximity to coastal waters, but which are substantially influenced by freshwater flows". Their high productivity provides habitats, refuge areas and food sources for a wide range of aquatic animals from resident brackish to freshwater and marine migratory species [4]. Transitional waters support important ecosystem services, including good water quality, fisheries, aquaculture and tourism, as well as agricultural activities in their watersheds [5]. Anthropic activities expose these ecosystems to continuous internal and external disturbance [2,6–8], including nitrogen (N) pollution arising from agricultural and urban activities, which poses potential threats to biodiversity and ecosystem functioning [3,9,10]. In addition, an increase in N-loads can significantly compromise water quality, promoting the development of micro and macroalgal blooms [11,12]. This, in turn, could alter the species composition and feeding behaviour of the aquatic animal community, from primary consumers to top predators. Changes in the availability and quality of basal food sources can affect the distribution of organisms and the feeding links between trophic levels, with effects on the stability and structure of the entire food chain [2,12,13]. Increased N-loads could thus also compromise, either directly or indirectly, the persistence of ecologically and economically important fish species [14].

In the Mediterranean area, the European eel, *Anguilla anguilla* (Linnaeus, 1758), and the annular seabream, *Diplodus annularis* (Linnaeus, 1758), are widespread and among the most important fishery resources [15,16]. However, in the last two decades, European eel populations have collapsed due to low recruitment and habitat alteration, and the species has been classified as 'critically endangered' since 2014, according to the International Union for Conservation of Nature [17]. It is known that both fish species are generally characterised by a high degree of omnivory and trophic plasticity depending on the composition and abundance of the available prey [18,19]. Specifically, the annular seabream, *Diplodus annularis*, is a demersal omnivorous species, feeding opportunistically on a wide variety of prey including zoobenthos, algae and plants. The European *Anguilla anguilla* is a generalist predator feeding mainly on invertebrates and fish but it also exhibits scavenger behaviour, feeding on dead animals including fish. These trophic traits can be expressed differently by individuals within the population [18,20,21]. Due to their omnivory, the trophic strategies of these species can directly reflect variations in the inputs determining the trophic status of the waters and thus the quality and availability of potential prey. Thus, understanding the patterns underlying the trophic choices of these fish species and their associated food webs is crucial for ecosystem management and the conservation of their habitats.

Several studies have been carried out on the diet of eels and seabream, often based on gut content analysis [19,22–25]. However, gut content analysis provides only a snapshot of a consumer diet, which is assumed to vary over time [7,26,27]. Furthermore, individuals often have no recognisable prey in their stomach, and description of the trophic links between species thus requires large samples [28].

Carbon ($\delta^{13}C$) and nitrogen ($\delta^{15}N$) stable isotope analysis is increasingly becoming useful tool for detecting organic and inorganic matter sources and understanding species' foraging behaviour and the relationships between organisms. It is thus useful for reconstructing food webs in aquatic ecosystems [7,29–31]. The isotopic ratio of these elements in consumer tissues reflects that of the assimilated food sources in a predictable way [7,32]. $\delta^{13}C$ signatures vary considerably among primary producers, generally with lower values in marine than terrestrial aquatic vegetation. This makes it possible to disentangle the contribution of various basal sources to food networks [7,31,33–36]. The $\delta^{15}N$ values gradually increase with each trophic level, thus providing information on the position of organisms in the food web [31,37,38]. In parallel, the $\delta^{15}N$ values of primary producers reflect the nature (organic or inorganic) and the source of nitrogen inputs (natural or anthropogenic) in a predictable way. $\delta^{15}N$ is thus also useful for tracking anthropogenic N pollution in water bodies and across trophic levels in food webs [11,39–41].

The main purpose of this study was to describe and analyse the diets and food webs of the eel *Anguilla anguilla* and the annular seabream *Diplodus annularis* in three neighbouring Mediterranean coastal lakes characterised by different eutrophication levels. It is known that energy flows and the transfer of nutrients depend primarily on the foraging choices of each organism within the community [31]. Similarly, the high trophic generalism and omnivory generally observed in *A. anguilla* and *D. annularis* [19,24,25,42,43] can be the result of different foraging strategies adopted by each individual within their respective populations.

In order to obtain highly detailed information and to consider variability in the use of resources by *A. anguilla* and *D. annularis*, the diet of the two species was obtained from trophic links of each individual within a population as determined by means of the Isotopic Trophic Unit (ITU) approach [31].

Isotopic Trophic Units are defined as groups of individuals with similar isotopic signatures occupying the same position in the $\delta^{13}C$-$\delta^{15}N$ niche space [31].

We studied the diet of each population in detail without excluding a priori any food source in the area. We hypothesised that differences in invertebrate and fish assemblages across lakes with differing trophic status could determine patterns of variation in the trophic niche width of the two fish species depending on their specific feeding habits. Specifically, we sought to verify whether a lower abundance and diversity of species at higher trophic levels caused *A. anguilla* to become more generalist and *D. annularis* to become more specialized.

2. Materials and Methods

2.1. Study Area

The samplings were carried out in three neighbouring Mediterranean brackish costal lakes located on the Tyrrhenian coast of central Italy (42°28′00″ North–12°51′00″ East): Lake Caprolace, Lake Fogliano and Lake Sabaudia (Figure 1). The three lakes respectively have a surface area of about 3 km^2, 4 km^2 and 3.9 km^2, and mean depths of 3 m, 2 m and 10 m. They are classified as non-tidal lagoons with a maximum tidal excursion of 0.21–0.23 m [44–46]. Salinity generally varies between 33.7 and 38.1 PSU in Caprolace, 29.9 and 39.2 PSU in Fogliano, and 28.8 and 33.7 PSU in Sabaudia. The annual average was 36.3 ± 0.8 PSU in Caprolace, 35.3 ± 0.8 PSU in Fogliano and 31.7 ± 0.9 PSU in Sabaudia in 2006–2010 [44]. Data are expressed as mean ± standard error.

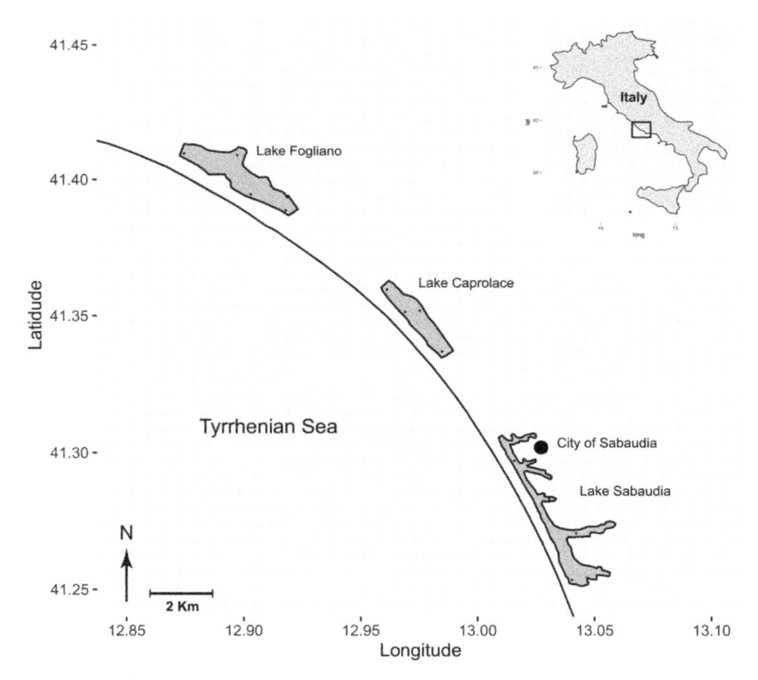

Figure 1. Map of the sampling area. The map shows the costal lakes of Caprolace (LP), Fogliano (IP) and Sabaudia (HP) located on the Tyrrhenian coast of central Italy (42°28′00″ North–12°51′00″ East).

The lakes are affected by various forms of anthropogenic disturbance related to organic and inorganic nitrogen inputs from urban treated sewage, livestock farming and agricultural activities, which are widespread in the surrounding areas [2,3].

On average, the mean concentration of total nitrogen was 383.6 ± 23.21 μg/L in Caprolace, 662.6 ± 66.70 μg/L in Fogliano and 1006.1 ± 49.97 μg/L in Sabaudia in 2006–2010. Santoro et al. [2] found the same trend in nitrate concentrations, with 12.2 ± 2.9 μg/L, 42.4 ± 61.3 μg/L and 91.9 ± 70.24 μg/L in Caprolace, Fogliano and Sabaudia respectively in the same period.

Lake Caprolace and Lake Fogliano (hereafter respectively LP and IP), characterised respectively by low and intermediate levels of eutrophication [3,47], are Sites of Community Importance (SCIs) located within the Circeo National Park (Lazio).

Lake Caprolace does not receive water inputs from the hinterland, while Lake Fogliano is affected by nutrient inputs from both the River Rio Martino and the livestock breeding activities practised in the surrounding areas. The annual concentration of *Chlorophyll a* was generally lower in Caprolace (2.1 ± 0.4 μg/L) than Fogliano (5.8 ± 1.2 μg/L) in 2006–2010.

Lake Sabaudia, the southernmost lake (hereafter HP), is affected by the highest anthropogenic pressure [3], mainly due to runoff from both the city of Sabaudia and cultivated fields in the surrounding areas as well as fishing and mussel farming. In this lake, freshwater inputs are present throughout the year.

Annual algal biomass and *Chlorophyll a* concentrations in this lake vary from 10.2 to 40.9 μg/L, with an average recorded value in 2006-2010 of 24.2 ± 6.15 μg/L. Further details regarding the study area can be found in Santoro et al. [2] and Jona-Lasinio et al. [3].

2.2. Field Collections

Samples of basal resources (primary producers and detritus), invertebrates and fish were collected in 4 sites per lake between April and May 2012, when primary productivity and invertebrate abundances were high. The sampling sites within each lake were selected from areas with heterogeneous physical and biotic characteristics and a range of anthropogenic impacts deriving from the surrounding areas [2,3]. The sampling sites were located at the northern and southern ends of each lake, and both on the landward and seaward sides (see also Santoro et al. [2]). Macrophytes, algae, and detritus samples were collected by hand and invertebrates by Van Veen grab (volume: 3.5 L) in three replicates per sampling site. The dominant macrophytes were *Ruppia* sp. and *Cymodocea nodosa* (Ucria) Ascherson, while the macroalgae were represented by taxa of the genera *Chetomorpha*, *Chondria*, *Gracilaria*, *Rytiphloea* and *Ulva*. The detritus was mostly composed of fragments of dead leaves delicately scraped to remove any epibionts and rinsed in distilled water. Phytoplankton samples were collected using a plankton net (20-μm mesh size) and concentrated by centrifugation (2000 rpm for 20 min).

Samples of fish were collected once a day for 3 days in each site. In order to collect pelagic, benthic, resident and migratory fish species, fish samples were collected using fixed weirs and fishing traps placed on the bottom. The fishing traps, made of very fine mesh (0.5 cm), were 1.5 m in diameter at the mouth and were composed of four consecutive chambers of decreasing diameter with a total length of 3.6 m. In addition to *A. anguilla* and *D. annularis*, the sampled fish community included the sand smelt *Atherina boyeri* (Risso, 1810), black goby *Gobius niger* (Linnaeus, 1758) and the mullets *Chelon ramada* (Risso, 1827), *C. aurata* (Risso, 1810), *C. saliens* (Risso, 1810) and *C. labrosus* (Risso, 1827), which are known to be prey species of *A. anguilla* and *D. annularis* [19,24,25,42,43]. Further fish samples included species belonging to the Sparidae, Scorpaenidae, Clupeidae, Cyprinodontidae, Blenniidae and Belonidae, Gobiidae, Labridae, Moronidae, Mugilidae, Soleidae and Syngnathidae families. Standard length measured in centimetres was recorded for each fish specimen. For each fish species, individuals of different sizes were collected in order to reduce the effects of size variability on isotopic signals. From the sampled fish specimens, including *A. anguilla* and *D. annularis*, samples of dorsal white muscle were taken. This tissue provides a long-term (several months) integrated indicator of food sources due to its slow turnover with respect to other tissues (e.g., liver and blood) [32].

After collection, all samples were transported to the laboratory, where specimens were sorted, counted, and identified to the lowest possible taxonomic level and processed for the stable isotope analysis.

2.3. Stable Isotope Analysis (SIA)

Samples were individually stored at −80 °C and freeze-dried for 24 h. Fish specimens were considered individually for isotopic analysis. Muscle samples were also taken from large invertebrates such as crustaceans, for which the tissue was taken from the claws, and bivalves and sea snails, whose tissue was taken from the feet [7]. When present, shells, valves and other exoskeletal parts of animals were removed under dissection microscopes in order to avoid tissue acidification before the stable isotope analysis. For small invertebrates (such as amphipods and polychaetes), the whole body was used. Samples were individually analysed. Plankton biomass was analysed as a whole due to the difficulty of obtaining sufficient biomass for isotopic analysis.

Before the stable isotope analysis, each sample was homogenised to a fine powder using a ball mill (Mini-Mill Fritsch Pulverisette 23: Fritsch Instruments, Idar-Oberstein, Germany). When necessary, samples were pre-acidified using 1M HCl according to the drop-by-drop method [48] in order to eliminate inorganic carbon and re-dried (60 °C) for 72 h to remove the remaining moisture. $\delta^{15}N$ signatures were measured in un-acidified powders to prevent acidification from interfering with the nitrogen analysis [30,49].

Aliquots of 0.25 ± 0.10 mg for the animals and 2.00 ± 0.10 mg for basal resources were placed into tin capsules for C and N stable isotopic analysis (SIA). Each sample was analysed in two replicates. The analyses were carried out using a continuous flow mass spectrometer (IsoPrime100, Isoprime Ltd., Cheadle Hulme, United Kingdom) coupled with an elemental analyser (Elementar Vario Micro-Cube, Elementar Analysensysteme GmbH, Germany).

The isotopic signatures of each sample were expressed in δ units ($\delta^{15}N$; $\delta^{13}C$) as parts per thousand (‰) deviations from international standards (atmospheric N_2 for N; PD-belemnite (PDB) carbonate for C), in accordance with the formula:

δX (‰) = [(Rsample − Rstandard)/Rstandard] × 10^3 [50], where X is ^{13}C or ^{15}N and R is the corresponding ratio of heavy to light isotope for the element ($^{13}C/^{12}C$ or $^{15}N/^{14}N$). Outputs were standardised with the internal laboratory standard Caffeine IAEA-600 ($C_8H_{10}N_4O_2$). Measurement errors were found to be typically smaller than 0.05‰.

2.4. Data Analysis

Differences between lakes in terms of animal community composition (considering both fish and benthic invertebrates) were tested using contingency tables based on chi-square (χ^2) tests, Monte Carlo permutation tests and the associated Cramer's V index (a measure of the strength of association among communities; Past 3.0 software package). Specimens collected in each sampling site (and replicates) were grouped by type or taxon (respectively for basal resources and animals) for each lake.

The Shannon diversity index (Hs) of invertebrate fauna for each lake was calculated at family level considering a total abundance of the taxa collected in each lake. Given that assessing Shannon diversity is only possible at the level of equal identification of all taxa, the few individuals belonging to the Gastropoda, Oligochaeta, Nematoda, and Nemertea classes (together accounting for less than 1.5% of total fauna) were excluded from the Shannon diversity index computation. Hutcheson's diversity *t*-test and the associated bootstrap procedure (9999 replicates), both available in the Past 3.0 software package, were applied to Hs values to test for significant differences [51]. Hutcheson's diversity *t*-test is a modified version of the classic *t*-test and is based on comparison of Hs variances. The *t* statistics of Hutcheson are defined as:

$$t = \frac{|Hs_i - Hs_j|}{\sqrt{var(Hs_i) + var(Hs_j)}} \quad (1)$$

which follows Student's *t* distribution. In the equation, *i* and *j* referred to the invertebrate communities of the lakes in paired comparisons, *Hs* represents the Shannon diversity index and *var(Hs)* its variance.

The isotopic values of collected organisms were used to reconstruct the diets of the eel *Anguilla anguilla* and the annular seabream *Diplodus annularis* in each lake. The diets were estimated on the basis of the Isotopic Trophic Unit (ITU) method [31]. The isotopic signatures of single basal resources, invertebrates, and fish were represented in the bi-dimensional isotopic space (Figure S1). This was subdivided into squares (ITUs) corresponding to 1 × 1‰ $\delta^{15}N$ and $\delta^{13}C$ values, starting from the lowest $\delta^{13}C$ value in the dataset and a $\delta^{15}N$ value of zero. The ITUs were thus identified and labelled (Figure S1).

The diets of each ITU containing individuals of the two fish species were calculated by means of Bayesian Mixing Models (R software ver. 3.5.3, SIMMr package) [52] considering a Trophic Enrichment factor (TEF) of 3.4 ± 1.0‰ for $\delta^{15}N$ and 1.0 ± 0.5‰ for $\delta^{13}C$ [18,37,49,53–56] and uninformative priors. These TEF values (expressed as mean ± standard deviation) are considered a robust and widely applicable assumption in the presence of multiple trophic pathways and different types of food sources [37,56]. For all SIMMr models, we ran three Markov Chain Monte Carlo chains of 300,000 iterations each with a burn-in of 200,000 and a thinning rate of 100 iterations. We assumed that all incoming food items had the same probability of being included in the consumer's diet. The model considers both variance in the isotopic signatures of the resources and uncertainty regarding the trophic enrichment of the consumer (TEF). The model results were expressed in the form of a probability distribution of plausible contribution values. The central tendency values of the distribution (mode, mean, median) allowed us to identify the most important food sources, while the upper and lower limits of the credibility ranges (CI: 50%, 75%, 95%) revealed the range of feasible contributions. The pool of food sources was selected based on the mixing model outputs in accordance with Rossi et al. [31]. Since the *A. anguilla* and *D. annularis* diet was obtained by starting from the foraging choice of each individual, the overall contribution of some food sources, important at individual level (>5%), could be relatively small (<5%) if considered at the population level (see also [31]). In order to obtain detailed information on the diet of the eel and the seabream these contributions were also considered.

Individuals other than *A. anguilla* and *D. annularis*, including basal resources and invertebrates, were excluded from ITU-consumers (but not from potential ITU food sources) before performing the Bayesian mixing models. This was done in order to correctly estimate the diet of *A. anguilla* and *D. annularis*. The set of potential ITU food sources was considered on the entire $\delta^{13}C$ axis and within a given range on the $\delta^{15}N$ axis, i.e., within ±3.4‰ (the TEF) of the value of the consumer [31]. The Bray–Curtis similarity index (BC), based on the contribution of each resource to the diet of the two fish, was also calculated in order to quantify the diet similarity among lakes [36,56]. BC is expressed as proportional similarity ranging from 0, when no common food sources are found for the compared groups, to 1, when the compared groups have the same food sources in the same proportions [36,56].

The symmetric overlap in resource use [57–59] was measured in accordance with the Pianka equation [59]:

$$O_{jk} = \frac{\sum_{i=1}^{n} p_{ij} p_{ik}}{\sqrt{\sum_{i=1}^{n} \left(p_{ij}\right)^2 \sum_{i=1}^{n} \left(p_{ik}\right)^2}} \tag{2}$$

where the Pianka index (O_{jk}) represents a symmetric measure of overlap between species j and k, and p_{ij} and p_{ik} are the proportional contributions of any given resource i used by species j and species k. The Pianka index ranges from 0 (overlap absent) to 1 (complete overlap).

Chesson's selectivity index [60] was calculated for each food item to determine possible preferences for particular food sources among those offered:

$$\alpha_i = \frac{r_i / n_i}{\sum_{i=1}^{m} (r_i / n_i)} \tag{3}$$

where α_i is Chesson's selectivity index, m is the number of food source types, r_i is the proportion of food type i in the diet and n_i is the proportion of food type i in the environment. The value of α_i ranges from 0 to 1, with 0 indicating complete avoidance, values above 1/m indicating preference and 1 indicating absolute preference [61]. Since consumer isotopic ratios provide an integrated measure of prey assimilated over time, we hypothesized that the composition of the taxon in each lake did not vary considerably over the course of a season. Therefore, the Chesson index based on the relationship between assimilated prey and its abundance in the environment could measure the selectivity of food products with a good approximation.

χ^2 tests were performed to test for differences between lakes in terms of the relative abundance of fauna and differences between food sources in terms of their proportional contribution to the diet of the fish population in each lake. Although it is not possible to establish a theoretical expected value, a χ^2 test was performed to test for possible differences between IP and LP and between HP and LP, considering the least polluted lake as the reference value.

Differences between lakes in both the δ^{15}N and δ^{13}C isotopic signatures of basal resources and fauna were tested by one-way ANOVA for comparisons between normal distributions (Shapiro–Wilk normality test, $p > 0.05$) while the Mann–Whitney with Bonferroni correction in cases of multiple comparisons and Kruskal–Wallis tests were used if non-normality was observed (Shapiro–Wilk normality test, p-value < 0.05). Levene's test for variances was used to test for differences within and between lakes in the δ^{13}C variance of primary producers. Kruskal–Wallis tests and associated Mann–Whitney pairwise comparisons were also used to compare the proportional contribution of food items to the diet of *A. anguilla* and *D. annularis*.

The niche metrics for both species in each lake were also calculated [62–64]. These metrics, originally proposed by Layman et al. [62] for application at community level, can be used at population level to obtain information about trophic diversity within a single population [35,63,64]. These included the ranges (highest to lowest) of δ^{13}C (Carbon Range, CR) and δ^{15}N (Nitrogen Range, NR) values. CR provides information about the variety of food sources exploited by the population (i.e., its trophic generalism), while NR indicates the number of trophic levels (i.e., degree of omnivory) of the population. The isotopic niche widths of both *A. Anguilla* and *D. annularis* were calculated as SEAc (Standard Ellipse Area corrected by degree of freedom) using R software ver. 3.5.3, SIBER analysis package [64,65]. The SEAc encompasses the core (about 40%) of the population's isotopic observations. This is a solid metric for comparing the isotopic niche of populations regardless of sample size and any isotopic outliers in the data [62,64]. Linkage density (L/S) was measured as the average number of feeding links (L) per ITU (S). Finally, based on the proportional contribution of each food source, the trophic niche width (TNW) of each population was measured as the diversity of resources consumed (Hs) by each population and compared among lakes. If not specified otherwise, the results are reported as mean ± standard error (s.e.).

3. Results

3.1. Community Composition and Isotopic Signatures

A total of 8752 samples comprising basal resources, invertebrates and fish were collected from the three lakes, 8645 (148 taxa) of which were invertebrates and fish (Table 1, Table S1).

Malacostraca (Amphipoda, Decapoda and Isopoda), Gastropoda, Anthozoa, Bivalvia, Polychaeta and Ophiuroidea together made up 93.09 ± 3.86 % of invertebrates.

Invertebrate abundance was lower in IP than the other two lakes (Table 1, paired-χ^2 test, χ^2 at least 20.46, p-value always <0.0001, Table S2). The composition of both the invertebrate and fish community also varied (contingency table, χ^2 at least 170.2, p always <0.001, Cramer's V at least 0.46, Tables S2 and S3). The abundance of some taxa, such as Decapoda and Anthozoa, decreased, while that of others (such as Amphipoda) increased with the pollution level of the lake (paired-χ^2 test, χ^2 at least 48.36, p-value always <0.0001). The number of fish taxa varied, i.e., 23 in LP, 17 in IP and 7 in HP. The relative abundance of fish differed between lakes (paired-χ^2 test, χ^2 at least 34.21 p-value always <0.0001) and was lowest in HP. The standard length of *Anguilla anguilla* was lower in HP (35.10 ± 3.63 cm) than the other two lakes (48.81 ± 6.07 cm in LP, 47.66 ± 2.12 cm in IP) (Mann–Whitney test with Bonferroni correction in cases of multiple comparisons, U = 13.0, p-value always <0.05). Similarly, *Diplodus annularis* had an average standard length of 6.25 ± 0.14 cm in HP, which was lower than LP

(8.76 ± 0.31 cm) and IP (9.43 ± 0.41 cm) (Mann–Whitney test with Bonferroni correction in cases of multiple comparisons, U = 2.0, p-value <0.05).

Table 1. Parameters describing the communities in each Lake. LP, IP and HP: low, intermediate and high eutrophication. N° indicates the sample size. Numbers in parentheses indicate the number of samples analysed. Community indicates both fish and benthic invertebrates. Stable isotopes of $\delta^{13}C$ and $\delta^{15}N$ are reported as mean (‰) ± s.e. For each parameter, different superscript letters (a,b,c) indicate differences between lakes (one-way ANOVA or Mann–Whitney test; $p < 0.05$).

	LP	IP	HP
N°			
Community	2942 (417)	2777 (502)	2926 (340)
Basal resources	28 (28)	51 (51)	28 (28)
Invertebrates	2793 (268) ^{a}	2526 (251) ^{b}	2848 (262) ^{a}
Fish	149 (145) ^{a}	251 (251) ^{b}	78 (78) ^{c}
$\delta^{13}C$ (‰)			
Community	−13.34 ± 0.17 ^{a}	−15.96 ± 0.13 ^{b}	−14.77 ± 0.15 ^{c}
Basal resources	−15.83 ± 0.78 ^{a}	−18.84 ± 0.80 ^{b}	−22.03 ± 1.04 ^{c}
Invertebrates	−12.65 ± 0.22 ^{a}	−15.63 ± 0.19 ^{b}	−14.42 ± 0.18 ^{c}
Fish	−14.67 ± 0.21 ^{a}	−16.30 ± 0.17 ^{b}	−16.09 ± 0.15 ^{b}
$\delta^{15}N$ (‰)			
Community	5.94 ± 0.16 ^{a}	8.36 ± 0.14 ^{b}	10.54 ± 0.14 ^{c}
Basal resources	3.65 ± 0.51 ^{a}	4.46 ± 0.38 ^{a}	6.70 ± 0.57 ^{b}
Invertebrates	4.42 ± 0.16 ^{a}	7.31 ± 0.21 ^{b}	9.82 ± 0.14 ^{c}
Fish	8.79 ± 0.17 ^{a}	9.41 ± 0.15 ^{b}	13.21 ± 0.16 ^{c}

Among the basal resources, detritus showed depleted $\delta^{13}C$ values, while primary producers were $\delta^{13}C$-enriched (Figure 2). $\delta^{13}C$-enrichment was also observed in pelagic fish with specialist diets such as *Atherina boyeri* ($\delta^{13}C$ = −15.34 ± 0.08‰ in LP, −17.45 ± 1.35‰ in IP and −16.14 ± 0.90‰ in HP). Since neither the mean nor the variance (σ^2) of $\delta^{13}C$ in the primary producers differed significantly either within each lake or between lakes (one-way ANOVA and associated Levene's test for homogeneity of variances, F at least 0.1412, p-value always >0.05), we concluded that the presence of a salinity gradient within a lake could not have an effect on the isotopic variability of the baseline. $\delta^{15}N$ values of primary producers increased with eutrophication (one-way ANOVA, F: 5.80, p <0.01).

The isotopic differences observed in basal resources reflected those observed in the whole community (Table 1, Figure 3; Kruskal–Wallis, Hc at least 127.1, p-value < 0.001; for $\delta^{13}C$ Mann–Whitney with Bonferroni correction for multiple comparisons, U at least 45,590, p-value always <0.001 and for $\delta^{15}N$ Mann–Whitney, U at least 421.57 with Bonferroni correction for multiple comparisons, p-value always <0.001, Figure 3) and in *A. anguilla* and *D. annularis.*

The $\delta^{13}C$ and $\delta^{15}N$ isotopic signatures of *Anguilla anguilla* and *Diplodus annularis* differed between lakes (Kruskal–Wallis, Hc at least 35.5, p-value <0.001). $\delta^{13}C$ values were higher in the least polluted lake (Table 2; Mann–Whitney with Bonferroni correction for multiple comparisons, U at least 3392.5, p-value always <0.001), while $\delta^{15}N$ values increased with eutrophication (Table 2; Mann–Whitney with Bonferroni correction for multiple comparisons, U at least 555, p-value always <0.001).

Specifically, in *A. anguilla,* the more generalist of the two species, $\delta^{13}C$ values reflected the shift of inputs from marine to terrestrial origin passing from the least to the most eutrophic lake (Figure 2 and Tables 1 and 2).

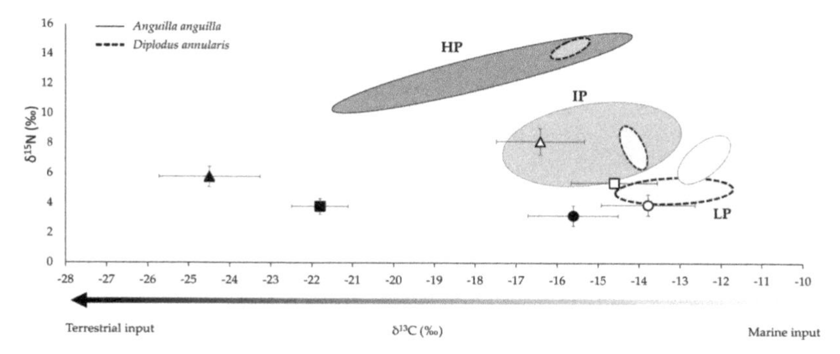

Figure 2. Isotopic standard ellipse areas (SEAcs) of *Anguilla anguilla* (continuous line) and *Diplodus annularis* (dashed line) in lakes with low (LP), intermediate (IP) and high (HP) eutrophication. Isotopic signatures (Mean ± s.e.) of primary producers (empty symbols) and detritus (full symbols) in lakes with low (circle), intermediate (square) and high (triangle) eutrophication. The greyscale reflects the origin of the main organic matter inputs from terrestrial (dark grey, left), to marine (light grey, right).

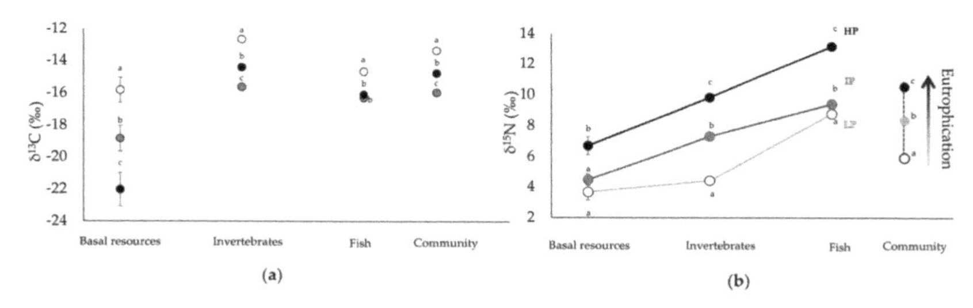

Figure 3. $\delta^{13}C$ **(a)** and $\delta^{15}N$ **(b)** values (‰) of basal resources, invertebrates, fish and the whole animal community in each lake. LP, IP and HP: low, intermediate and high eutrophication. Isotopic values are reported as mean ± s.e. Greyscale indicate degrees of eutrophication: LP (white), IP (grey), HP (black). Arrow indicates increasingly eutrophic conditions. Different letters (a, b, c) within panels indicate differences between lakes (Mann–Whitney test with Bonferroni correction for multiple comparisons; *p*-value <0.05).

Table 2. Isotopic niche and food web metrics of the eel *Anguilla anguilla* and the seabream *Diplodus annularis* in each lake. LP: low, IP: intermediate, HP: high eutrophication. N: sample size, ITUs: Isotopic Trophic Units, $\delta^{13}C$ (‰) and $\delta^{15}N$ (‰) (mean ± s.e.), CR: Carbon Range, NR: Nitrogen Range, L: number of feeding links, S: number of ITUs in the diet, L/S: Linkage density, SEAc: Standard Ellipse Area "corrected" (SEAc) by degree of freedom, TNW: Trophic Niche Width. For details of metrics, please refer to the materials and methods section. For each parameter, different superscript letters (a,b,c) indicate differences between lakes (Mann–Whitney test with Bonferroni correction for multiple comparisons; *p*-value <0.05).

	Anguilla anguilla			*Diplodus annularis*		
	LP	IP	HP	LP	IP	HP
N	8	16	10	8	6	8
ITUs						
$\delta^{13}C$ (‰)	-12.76 ± 0.11^{a}	-15.07 ± 0.15^{b}	-16.31 ± 0.16^{c}	-13.49 ± 0.16^{a}	-16.23 ± 0.14^{a}	-15.34 ± 0.19^{b}
$\delta^{15}N$ (‰)	8.27 ± 0.19^{a}	9.62 ± 0.12^{a}	13.58 ± 0.11^{b}	5.81 ± 0.15^{a}	7.99 ± 0.14^{b}	10.24 ± 0.15^{c}
CR	2.05	9.52	6.19	2.17	0.35	0.65
NR	3.15	5.79	3.51	0.67	0.96	0.74
Taxa						

Table 2. *Cont.*

	Anguilla anguilla			*Diplodus annularis*		
	LP	IP	HP	LP	IP	HP
$\delta^{13}C$ (‰)	−12.29 ± 0.19 [a]	−15.01 ± 0.53 [b]	−17.20 ± 1.12 [c]	−13.00 ± 0.49 [a]	−14.00 ± 0.10 [a]	−15.57 ± 0.16 [b]
$\delta^{15}N$ (‰)	9.08 ± 0.32 [a]	9.50 ± 0.36 [a]	12.96 ± 0.51 [b]	7.73 ± 0.17 [a]	9.59 ± 0.29 [b]	13.87 ± 0.16 [c]
CR	1.55	4.66	6.19	2.17	0.35	0.65
NR	2.54	2.59	2.86	0.67	0.96	0.74
L	30	84	33	28	13	14
S	19	42	22	21	10	11
L/S	1.6	2.0	1.5	1.3	1.3	1.3
SEAc	1.46	9.62	4.84	1.55	0.45	0.32
TNW	1.81	2.06	2.32	2.15	2.28	1.98

3.2. Niche Metrics and Diet of *Anguilla anguilla*

The isotopic signatures and niche metrics of *Anguilla anguilla* varied among lakes (Table 2, Figures 2–4; Kruskal–Wallis, Hc least 12.06, *p*-value <0.001). The highest $\delta^{15}N$ values were observed in HP (Table 2; Mann–Whitney with Bonferroni correction in cases of multiple comparisons, U at least 0.1, *p*-value always <0.001). The Carbon Range increased with eutrophication (Figure 4, Table 2) and the largest Nitrogen Range was observed in the eutrophic lake (Table 2).

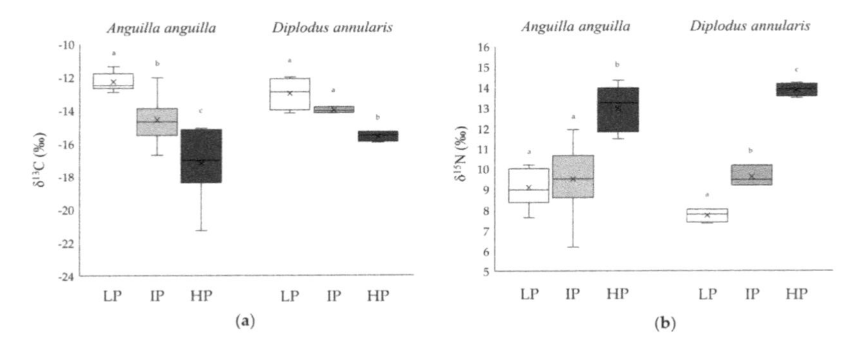

Figure 4. Boxplot of the distribution of $\delta^{13}C$ (**a**) and $\delta^{15}N$ (**b**) isotopic signature of *Anguilla anguilla* and *Diplodus annularis* in each lake: LP, IP and HP: low, intermediate and high eutrophication. For each lake, the thick horizontal line represents the median of the distribution, the box includes 50% of the data, the symbol (x) represents the mean and the whiskers reach the highest and lowest value within 95% of the distribution. Different letter (a, b, c) within panels indicates differences among lakes (Mann–Whitney test with Bonferroni correction in cases of multiple comparisons; $p < 0.05$).

Overall, no correlation between the body length and $\delta^{13}C$ (‰) of *Anguilla anguilla* was observed in any lake (Pearson correlation, $p > 0.05$).

Anguilla anguilla had 5 ITUs in LP and HP and 12 ITUs in IP, where the eel-resource ITU linkage density was highest (Table 2). ITU-based mixing models showed no differences between lakes in terms of the overall contribution of invertebrates to the eels' diet (Figure 5a).

By contrast, the consumption of basal resources increased and piscivory decreased with increasing levels of pollution (i.e., from LP to HP; Table 3 Figure 5a).

A. anguilla showed a generalist diet including 20 different categories of food source (Table 3, Figures 6 and 7). Some of these were common to the three lake populations (e.g., Actinopterygii, Bivalvia, Gastropoda, Decapoda and Polychaeta) but their consumption varied. The Bray–Curtis index (BC) applied to diet showed a lower similarity between the HP population and the others (76% similarity between LP and IP vs. 41% between LP and HP, and 54% between IP and HP). Specifically, in LP the diet of *A. anguilla* was mostly based on Actinopterygii (34.76% ± 1.90), Decapoda (27.84% ± 4.60) and Gastropoda (13.06% ± 0.40), in IP on Actinopterygii (30.65% ± 0.50) and Decapoda (26.13% ±

1.90) and in HP on Polychaeta (30.41% ± 0.60), Actinopterygii (12.18% ± 1.50), Bivalvia (10.35% ± 0.90), detritus (9.03% ± 0.50) and Decapoda (9.11% ± 1.90) (Table 3, Figure 6).

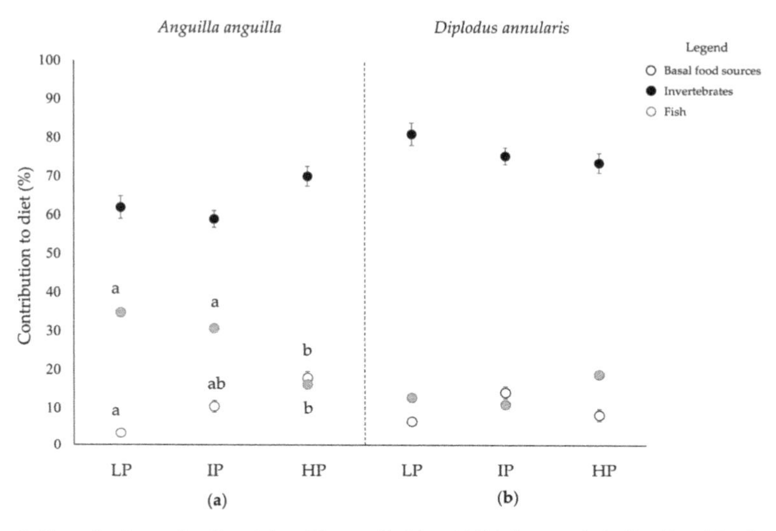

Figure 5. Contribution to the diet of *Anguilla anguilla* (**a**) and *Diplodus annularis* (**b**) of basal food sources (white), invertebrates (black) and fish (grey) in the lakes with low (LP), intermediate (IP) and high (HP) eutrophication. The overall contribution of basal resources, invertebrates and fish is reported as the mean (%) ± s.e. Different letters (a,b,c) within panels indicate differences between lakes in the contribution of food sources to the diet (χ^2-test, p-value <0.001).

Table 3. Proportional contribution (in %) of food sources to the diet of *A. anguilla* in each lake, obtained from ITU-based mixing models. LP: low anthropogenic pressure, IP: intermediate anthropogenic pressure, HP: high anthropogenic pressure. The contribution of each food source is reported as the mean (±s.e.). "Taxa" indicates the number of taxa belonging to the respective group in the diet of *A. anguilla*. The overall contribution of basal resources, invertebrates and fish is reported as the mean (%) ± s.e. Different superscript letters (a,b,c) indicate differences between lakes in the contribution of categories of food sources to the diet (χ^2-test, p-value <0.05). For details please refer to the methods section.

	LP		IP		HP	
Food Sources	**Taxa**	**Contribution**	**Taxa**	**Contribution**	**Taxa**	**Contribution**
TELEOSTS						
Actinopterygii	7	34.76 ± 1.90	12	30.65 ± 0.50	3	12.18 ± 1.50
CNIDARIANS						
Anthozoa	2	4.40 ± 0.60	4	3.17 ± 0.10	2	3.39 ± 0.20
Hydrozoa	-	-	-	-	1	1.67 ± 0.10
ASCIDIANS						
Ascidiacea	-	-	-	-	1	1.82 ± 0.10
BASAL RESOURCES						
Algae	1	0.75 ± 0.10	-	-	1	1.55 ± 0.10
Detritus	2	1.81 ± 0.40	4	5.33 ± 0.60	2	9.03 ± 0.50
Phytoplankton	-	-	1	0.42 ± 0.10	1	3.10 ± 0.10
Aquatic plants	1	0.75 ± 0.10	4	4.68 ± 0.40	1	4.19 ± 0.10
MOLLUSCS						
Bivalvia	1	0.43 ± 0.10	4	5.68 ± 0.40	4	10.35 ± 0.90
Gastropoda	8	13.06 ± 0.40	3	1.57 ± 0.20	2	2.95 ± 1.00
ANELLIDA						
Clitellata (Oligochaeta)	-	-	-	-	1	2.60 ± 0.10
Polychaeta	6	5.62 ± 0.20	5	6.99 ± 0.20	13	30.41 ± 0.60

Table 3. *Cont.*

Food Sources	LP		IP		HP	
	Taxa	Contribution	Taxa	Contribution	Taxa	Contribution
ECHINODERMS						
Eleutherozoa (Asteroidea)	1	0.29 ± 0.10	-	-	-	-
Euechinoidea (Echinoidea)	1	4.40 ± 0.10	-	-	-	-
Ophiuroidea	-	-	1	2.01 ± 0.10	-	-
ARTHROPODS						
Insecta	-	-	1	1.42 ± 0.10	1	1.62 ± 0.10
Malacostraca						
Amphipoda	4	4.74 ± 0.30	4	5.10 ± 0.40	2	4.13 ± 0.50
Decapoda	4	27.84 ± 4.60	7	26.13 ± 1.90	2	9.11 ± 1.90
Isopoda	2	1.15 ± 1.70	4	5.78 ± 0.30	-	-
NEMERTEANS						
Nemertea	-	-	1	1.08 ± 0.10	1	1.91 ± 0.10
BASAL RESOURCES		3.31 ± 0.35 ^{a}		10.43 ± 1.54 ^{ab}		17.87 ± 1.60 ^{b}
INVERTEBRATES		61.93 ± 2.92		58.93 ± 2.15		69.96 ± 2.56
FISH		34.76 ± 1.90 ^{a}		30.65 ± 0.50 ^{a}		16.18 ± 1.50 ^{b}

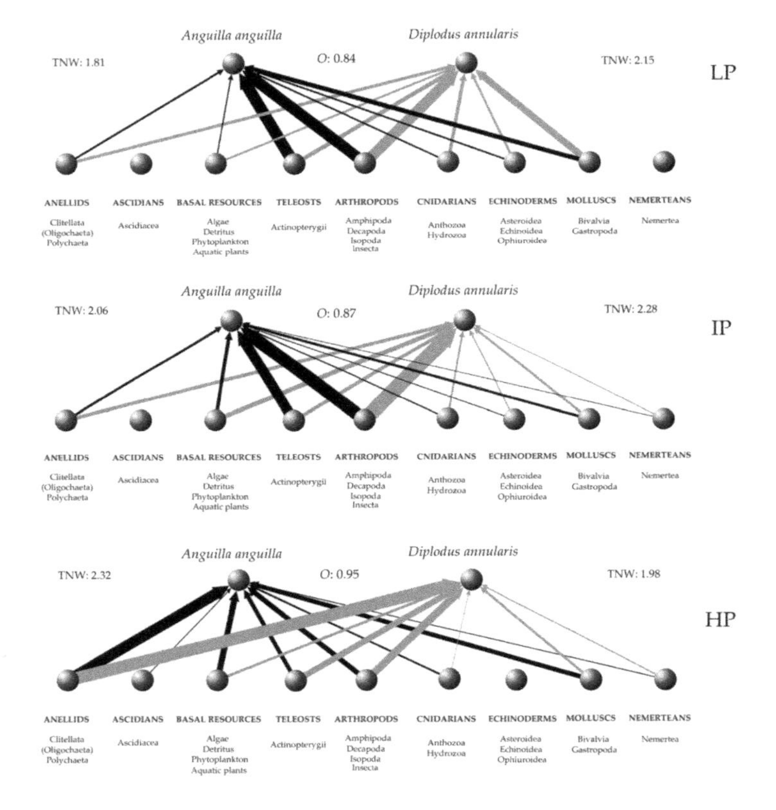

Figure 6. *Anguilla anguilla* and *Diplodus annularis* food webs in lakes with low (LP), intermediate (IP) and high (HP) eutrophication. Each node at the base of the food web represents a food source (in terms of class and respective families). Arrows point from each food item to its consumer: *Anguilla anguilla* (black arrows) and *Diplodus annularis* (grey arrows). The arrows' thickness is proportional to the trophic interaction strength. TNW indicates trophic niche width. *O* indicates the niche overlap between *D. annularis* and *A. anguilla*. For details of metrics, please refer to the results section.

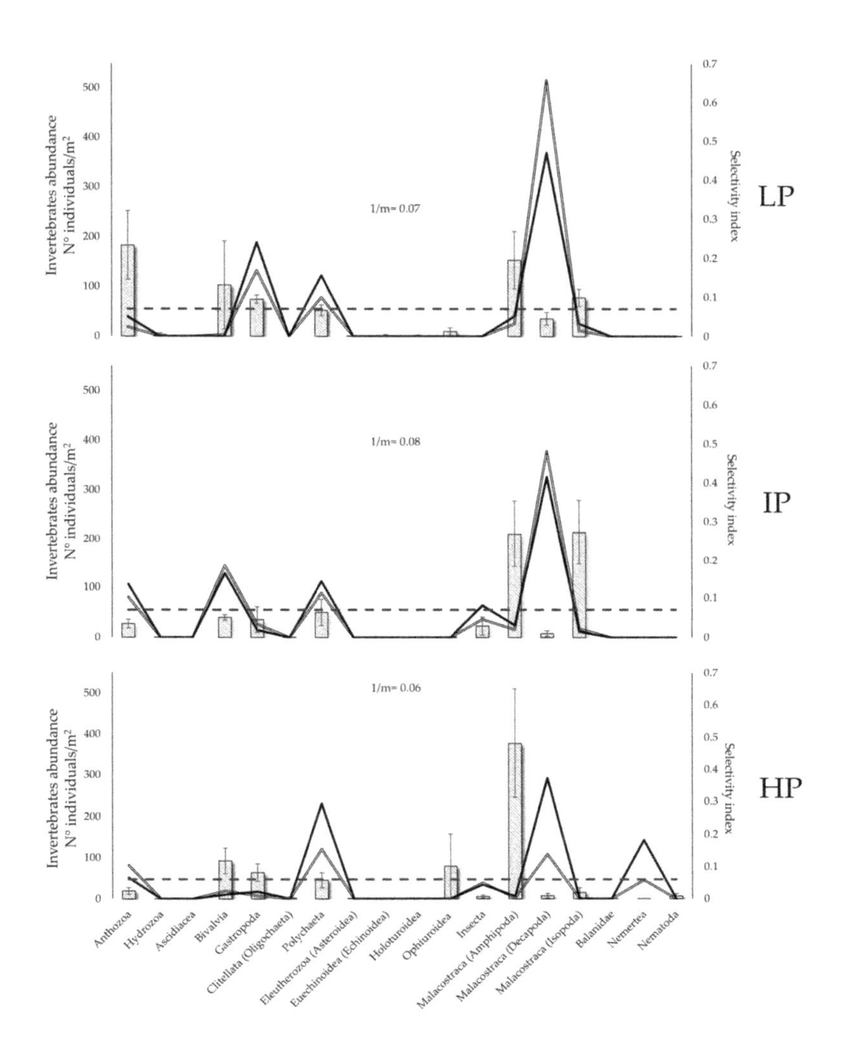

Figure 7. Invertebrate abundance in the environment (histograms), and selectivity values for each invertebrate group in the diet of *Anguilla anguilla* (double line) and *Diplodus annularis* (single thick line) in LP, IP and HP (low, intermediate and high eutrophication). Selectivity values greater than 1/m (dotted line) indicate preference.

The difference in resource use was associated with a difference in trophic niche width (Figures 2–6), which increased with eutrophication (TNW: 1.81, 2.06 and 2.32 in LP, IP and HP respectively), with significant differences between HP and LP (bootstrap comparison among populations, $p < 0.0001$).

3.3. Niche Metrics and Diet of *Diplodus annularis*

The $\delta^{13}C$ and $\delta^{15}N$ of *Diplodus annularis* varied across lakes (Figures 2–4, Kruskal–Wallis for both $\delta^{13}C$ and $\delta^{15}N$, Hc at least 35.5, p-value <0.001). In addition, no significant correlation between the body length and $\delta^{13}C$ (‰) of *Diplodus annularis* was observed in any lake (Pearson correlation, $p > 0.05$).

$\delta^{15}N$ increased with the level of pollution (Table 2, Figure 4). The Carbon Range was highest in LP while no differences in Nitrogen Range were observed between lakes (Table 2). The seabreams had four ITUs in LP and two ITUs in both IP and HP (Table 2). No differences in the linkage density (L/S) of the ITUs or in SEAc were observed between lakes (Table 2).

Mixing models based on the single ITU values showed similar average contributions of basal resources, invertebrates and fish in the lakes (Figure 5b; paired-χ^2 test, p-value always >0.05).

However, when taxa in each category were distinguished, diet similarity between LP and HP was 55%, while between IP and both LP and HP it was 61% (Bray–Curtis index, BC).

Overall, the diet of *D. annularis* was based on 17 different taxa, and invertebrates represented more than 70% of it in all lakes (Table 4, Figures 5 and 6). Among these, Decapoda (22.03% ± 1.37), Gastropoda (20.65% ± 1.06) and Actinopterygii (12.69% ± 0.49) contributed most to the diet of *D. annularis* in LP (Figure 6); Decapoda (27.86% ± 1.69), Actinopterygii (10.84% ± 0.28) and Amphipoda (9.86% ± 1.54) in IP (Figure 5); and Polychaeta (34.76% ± 1.12), Actinopterygii (18.56% ± 0.78) and Decapoda (14.67% ± 1.03) in HP.

Table 4. Diet composition of *Diplodus annularis* in each lake. Proportional contribution (in %) of food sources to the diet of *D. annularis* obtained from ITU-based mixing models. LP: low anthropogenic pressure, IP: intermediate anthropogenic pressure, HP: high anthropogenic pressure. The contribution of each food source is reported as the mean (± s.e.). "Taxa" indicates the number of taxa belonging to the respective group in the diet of *D. annularis*. The overall contribution of basal resources, invertebrates and fish is reported as the mean (%) ± s.e.

	LP		IP		HP	
Food Sources	**Taxa**	**Contribution**	**Taxa**	**Contribution**	**Taxa**	**Contribution**
TELEOSTS						
Actinopterygii	4	12.69 ± 0.49	5	10.84 ± 0.28	3	18.56 ± 0.78
CNIDARIANS						
Anthozoa	2	10.21 ± 1.31	3	5.15 ± 0.69	1	1.28 ± 4.75
Hydrozoa	-	-	-	-	-	-
BASAL RESOURCES						
Algae	1	2.01 ± 0.10	-	-	-	-
Detritus	1	1.56 ± 0.58	2	5.91 ± 0.29	1	3.02 ± 5.00
Phytoplankton	-	-	-	-	-	-
Aquatic plants	2	2.88 ± 0.91	4	8.07 ± 0.10	1	5.05 ± 0.28
MOLLUSCS						
Bivalvia	-	-	2	6.24 ± 0.76	2	3.37 ± 0.44
Gastropoda	6	20.65 ± 1.06	1	1.08 ± 5.00	2	4.29 ± 1.23
ANELLIDS						
Clitellata (Oligochaeta)	-	-	-	-	1	2.73 ± 5.00
Polychaeta	8	9.71 ± 0.55	3	10.89 ± 0.69	8	34.76 ± 1.12
ECHINODERMS						
Eleutherozoa (Asteroidea)	1	0.99 ± 5.00	-	-	-	-
Euechinoidea (Echinoidea)	1	6.04 ± 5.00	-	-	-	-
Ophiuroidea	-	-	1	3.54 ± 0.75	-	-
ARTHROPODS						
Insecta	-	-	1	3.1 ± 1.06	1	0.87 ± 5.00
Malacostraca						
Amphipoda	3	8.43 ± 1.13	3	9.86 ± 1.54	2	7.8 ± 1.10
Decapoda	5	22.03 ± 1.37	5	27.86 ± 1.69	2	14.67 ± 1.03
Isopoda	3	2.8 ± 0.31	4	4.85 ± 0.20	-	-
NEMERTEANS						
Nemertea	-	-	1	2.61 ± 0.71	1	3.6 ± 0.74
BASAL RESOURCES		6.45 ± 0.39		13.98 ± 1.08		8.07 ± 1.02
INVERTEBRATES		80.86 ± 2.7		75.18 ± 2.46		73.37 ± 3.61
FISH		12.68 ± 0.49		10.82 ± 0.28		18.57 ± 0.78

The trophic niche was significantly narrower in HP than LP and IP (Figure 6; TNW: 1.98 vs. 2.15 and 2.28 respectively; bootstrap comparison among populations, p <0.05). The niche overlap between *D. annularis* and *A. anguilla* decreased from the eutrophic to the unpolluted lake (O = 0.95, 0.87 and 0.84 in HP, IP and LP respectively).

4. Discussion

Our results indicate that anthropogenic inputs affected the composition and abundance of the lake animal community. Specifically, the diversity of fish decreased, and trophic choices of the eel *Anguilla anguilla* and the seabream *Diplodus annularis* changed, with increasing eutrophication. *A. anguilla*, which was more piscivorous than the seabream at low and intermediate eutrophication, increased its consumption of invertebrates and basal resources. On the other hand, the seabream, which fed more on invertebrates, increased its preference for polychaetes. The high selectivity for polychaetes in the highly eutrophicated lake could be due to the facilitated capture of these preys and their good contribution to the energy supply of their predators compared to other aquatic invertebrates [66–68]. Changes in the feeding choices of the two fish species resulted in increased interspecific niche overlap, suggesting that eutrophication may have strong bottom-up effects on interspecific interactions [69,70].

Previous research demonstrates that the δ^{15}N values of the aquatic biota reflect anthropogenic nitrogen inputs from the surrounding terrestrial areas [7,10,71,72]. In our study, increased eutrophication was associated with higher δ^{15}N in the fish community, as previously observed by Santoro et al. [2] and Jona-Lasinio et al. [3] for invertebrates and primary producers respectively. Our results indicate that in transitional waters, individual fish δ^{15}N could be a useful indicator of anthropogenic N transfer along food webs [10,73], while the range of δ^{13}C in the fish population could reflect the diversity of C inputs, emphasising the need for ecological monitoring in these productive ecosystems.

The distinct carbon isotopic signatures (δ^{13}C) of primary producers (e.g., terrestrial vs. aquatic vegetation) allowed us to discern the origin of the organic matter contributing to the nutrient pool of water bodies [7,33,34,74]. Specifically, depleted δ^{13}C organic matter values indicated the contribution to the organic matter pool of allochthonous (terrestrial) carbon, while enriched δ^{13}C indicated autochthonous primary production, as also observed in other aquatic ecosystems [7,34,35,74]. Although it was not in the remit of this study to investigate the cause, carbon enrichment was also visible in pelagic and strictly specialist species such as the sand smelt, *A. boyeri*. In addition, neither the mean nor the variance of δ^{13}C of primary producers differed significantly between sites within the same lake or between different lakes. This allowed us to exclude possible interferences in the isotopic baseline arising from environmental parameters such as the salinity and oxygenation levels of the respective area. The larger contribution of terrestrial organic matter in HP may be due to the large input of fresh water that this lake receives from the hinterland, as indicated by the low salinity generally observed in this lake compared to the other two. These results are consistent with what has been observed in similar environments [7,74,75].

Large Carbon Ranges suggest multiple carbon sources at the base of the food web [7,35,62–64]. The supply of organic matter from multiple sources in the three lakes was evident in the δ^{13}C values of *A. anguilla*, which shifted from marine to terrestrial input with the increasing pollution. Its larger Carbon Range in the eutrophic lake indicates that this species integrated both autochthonous and allochthonous carbon pathways, while in the unpolluted lake it relied mainly on the autochthonous one.

It is acknowledged that increased N-loads promote significant changes in aquatic productivity [76] that could potentially affect the composition of the prey community [77,78]. This in turn might be reflected in the feeding behaviour of consumers at all trophic levels [2,7,79].

In our study, increased N pollution resulted in altered community composition, with decreasing diversity, which seemed to affect the feeding preference and niche width of *Anguilla anguilla* and *Diplodus annularis*. The feeding regime of the two species is known to be characterised by marked generalism and trophic plasticity [19,22,42,43,80]. However, while the eel enlarged its trophic niche, feeding off multiple resources at various trophic levels in eutrophic conditions, the seabream concentrated on a small number of invertebrates. Although dietary changes are known to depend primarily on size and growth stage [81–83], in our study neither *A. anguilla* nor *D. annularis* showed a relationship between body size and δ^{13}C signatures. The greater trophic generalism and omnivory of eels with eutrophication may be due to the different density, accessibility and availability of the prey at higher trophic levels [19,22,24,42,43] as well as to an altered presence of potential competitors for the same food sources [78]. This was also confirmed by the results of the Bayesian mixing models, which showed a shift with eutrophication in the diet of *Anguilla anguilla* mostly from pelagic (fish) to benthic (invertebrates) prey, regardless of their abundance. Bouchereau et al. [24,25] reported Teleosts, amphipods and decapods as the predominant prey in the diet of *A. anguilla* in two North-Mediterranean lagoons, and that prey selection could be linked to the activity and accessibility of the prey itself. Rosecchi [42], Pita et al. [23] and Chaouch et al. [19] indicated molluscs (bivalves and gastropods), crustaceans, polychaetes and Teleosts as the main items in the diet of *D. annularis* in lagoons and coastal waters. Lammens et al. [84] and Dörner et al. [80] identified *A. anguilla* as belonging to the piscivorous community in many European lakes. In our study, the adoption of piscivory in the least eutrophic lake allowed the eel to reduce niche overlap [80,84,85] and therefore potential competition with other fish.

However, the trophic behaviour of the two species can be expressed differently by individuals within populations [18,20,21], enabling *A. anguilla* and *D. annularis* to include several food sources in their diet even in a single area [19,25]. In this context, the individual isotopic characterisations of carbon ($\delta^{13}C$) and nitrogen ($\delta^{15}N$), coupled with diet reconstruction at the isotopic trophic unit level (ITU, as recently proposed by Rossi et al. [31]) were crucial to the detailed estimation of the diets of the two fish species with high trophic generalism and omnivory. This allowed us to consider the possibility that each individual consumer could draw on the whole spectrum of potential food sources available within each lake [31]. In this way, we are able to describe the trophic plasticity and generalism of two ecologically and economic important fish species, and hence the real variation of the diet within the same populations under a range of eutrophication conditions.

5. Concluding Remarks

Understanding the trophic response of the community to eutrophication and depicting the structure of food webs in coastal lakes is still problematic [14], mainly due to the extraordinary biological diversity and complexity of the potential trophic links between species [2,86,87]. In our study, the stable isotope analysis of carbon ($\delta^{13}C$) and nitrogen ($\delta^{15}N$) provided an effective approach with which to (1) track the propagation of anthropogenic nutrient inputs along food chains, (2) evaluate the relative contributions of food sources to fish diets, and (3) quantify the trophic relationships between organisms [31,37,62,87–89]. Here, the diet resulting from the application of the method recently proposed by Rossi et al. [31] substantially improved our ability to understand the response of communities to increasing eutrophication, as well as its effect on the feeding behaviour and food choices of important fish species in Mediterranean coastal lakes. Together, our results confirm food web theory as a powerful approach for obtaining valuable information for the management and conservation of these complex and productive ecosystems.

Supplementary Materials: The following are available online at http://www.mdpi.com/2076-3417/10/8/2756/s1, Figure S1: Isotopic niche biplots of the community in each lake, Table S1: Carbon (δ13C) and Nitrogen (δ15N) Isotopic signature and abundances (n° Ind./m2) of invertebrate in each lake, Table S2: Contingency table for Invertebrates (Total Individuals) community composition in each lake, Table S3: Contingency table for fish (Total individuals) community composition in each lake.

Author Contributions: Conceptualisation, Supervision and Funding Acquisition: L.R. and M.L.C.; Field sampling: E.C. and G.C.; Investigation: S.S.C., E.C., G.C., F.F. and D.M.; Formal analysis and Visualisation: S.S.C. and G.C.; Writing—Original Draft Preparation: L.R., M.L.C., S.S.C.; Writing—Review & Editing: All Authors. All authors have read and agreed to the published version of the manuscript.

Funding: This research was funded by: SAMOBIS-project granted by Provincia Latina, PNRA-2015/AZ1.01 (M.L. Costantini) and PNRA16_00291 (L. Rossi) granted by MIUR-PNRA and the APC was funded by Progetti di Ricerca di Ateneo-RM11916B88AD5D75 (E. Calizza).

Acknowledgments: This work was supported by Latina Provincial Administration (Project: SAMOBIS), PNRA-2015/AZ1.01 (M.L. Costantini) and PNRA16_00291 (L. Rossi). We thank Mr. George Metcalf for revising the English text.

Conflicts of Interest: The authors declare no competing interests.

References

1. Basset, A.; Sabetta, L.; Fonnesu, A.; Mouillot, D.; Chi, T.D.; Viaroli, P.; Giordani, G.; Reizopoulou, S.; Abbiati, M.; Carrada, G.C. Typology in Mediterranean transitional waters: New challenges and perspectives. *Aquat. Conserv. Mar. Freshw. Ecosyst.* **2006**, *16*, 441–455. [CrossRef]
2. Santoro, R.; Bentivoglio, F.; Carlino, P.; Calizza, E.; Costantini, M.L.; Rossi, L. Sensitivity of food webs to nitrogen pollution: A study of three transitional water ecosystems embedded in agricultural landscapes. *Transit. Waters Bull.* **2014**, *8*, 84–97.
3. Jona-Lasinio, G.; Costantini, M.L.; Calizza, E.; Pollice, A.; Bentivoglio, F.; Orlandi, L.; Careddu, G.; Rossi, L. Stable isotope-based statistical tools as ecological indicator of pollution sources in Mediterranean transitional water ecosystems. *Ecol. Indic.* **2015**, *55*, 23–31. [CrossRef]
4. Elliott, M.; Whitfield, A.K.; Potter, I.C.; Blaber, S.J.M.; Cyrus, D.P.; Nordlie, F.G.; Harrison, T.D. The guild approach to categorizing estuarine fish assemblages: A global review. *Fish Fish.* **2007**, *8*, 241–268. [CrossRef]
5. Chapman, P.M. Management of coastal lagoons under climate change. *Estuar. Coast. Shelf Sci.* **2012**, *110*, 32–35. [CrossRef]
6. Orlandi, L.; Bentivoglio, F.; Carlino, P.; Calizza, E.; Rossi, D.; Costantini, M.L.; Rossi, L. δ15N variation in *Ulva lactuca* as a proxy for anthropogenic nitrogen inputs in coastal areas of Gulf of Gaeta (Mediterranean Sea). *Mar. Pollut. Bull.* **2014**, *84*, 76–82. [CrossRef]
7. Careddu, G.; Costantini, M.L.; Calizza, E.; Carlino, P.; Bentivoglio, F.; Orlandi, L.; Rossi, L. Effects of terrestrial input on macrobenthic food webs of coastal sea are detected by stable isotope analysis in Gaeta Gulf. *Estuar. Coast. Shelf Sci.* **2015**, *154*, 158–168. [CrossRef]
8. Fiorentino, F.; Cicala, D.; Careddu, G.; Calizza, E.; Jona-Lasinio, G.; Rossi, L.; Costantini, M.L. Epilithon δ^{15}N signatures indicate the origins of nitrogen loading and its seasonal dynamics in a volcanic Lake. *Ecol. Indic.* **2017**, *79*, 19–27. [CrossRef]
9. Howarth, R.W. Coastal nitrogen pollution: A review of sources and trends globally and regionally. *Harmful Algae* **2008**, *8*, 14–20. [CrossRef]
10. Calizza, E.; Favero, F.; Rossi, D.; Careddu, G.; Fiorentino, F.; Sporta Caputi, S.; Rossi, L.; Costantini, M.L. Isotopic biomonitoring of N pollution in rivers embedded in complex human landscapes. *Sci. Total Environ.* **2020**, *706*, 136081. [CrossRef]
11. Fox, S.E.; Teichberg, M.; Olsen, Y.S.; Heffner, L.; Valiela, I. Restructuring of benthic communities in eutrophic estuaries: Lower abundance of prey leads to trophic shifts from omnivory to grazing. *Mar. Ecol. Prog. Ser.* **2009**, *380*, 43–57. [CrossRef]
12. Martínez-Durazo, A.; García-Hernández, J.; Páez-Osuna, F.; Soto-Jiménez, M.F.; Jara-Marini, M.E. The influence of anthropogenic organic matter and nutrient inputs on the food web structure in a coastal lagoon receiving agriculture and shrimp farming effluents. *Sci. Total Environ.* **2019**, *664*, 635–646. [CrossRef]
13. Calizza, E.; Costantini, M.L.; Rossi, L. Effect of multiple disturbances on food web vulnerability to biodiversity loss in detritus-based systems. *Ecosphere* **2015**, *6*, 1–20. [CrossRef]
14. Carlier, A.; Riera, P.; Amouroux, J.-M.; Bodiou, J.-Y.; Desmalades, M.; Grémare, A. Food web structure of two Mediterranean lagoons under varying degree of eutrophication. *J. Sea Res.* **2008**, *60*, 264–275. [CrossRef]
15. Capoccioni, F.; Costa, C.; Canali, E.; Aguzzi, J.; Antonucci, F.; Ragonese, S.; Bianchini, M.L. The potential reproductive contribution of Mediterranean migrating eels to the *Anguilla anguilla* stock. *Sci. Rep.* **2014**, *4*, 1–7. [CrossRef]
16. Trojette, M.; Faleh, A.B.; Fatnassi, M.; Marsaoui, B.; Mahouachi, H.; El, N.; Chalh, A.; Quignard, J.P.; Trabelsi, M. Stock discrimination of two insular populations of *Diplodus annularis* (ACTINOPTERYGII: PERCIFORMES: SPARIDAE) along the coast of Tunisia by analysis of otolith shape. *Acta Ichtyologica et Piscicatoria* **2015**, *45*, 363–372. [CrossRef]
17. Bilotta, G.S.; Sibley, P.; Hateley, J.; Don, A. The decline of the European eel *Anguilla anguilla*: Quantifying and managing escapement to support conservation. *J. Fish Biol.* **2011**, *78*, 23–38. [CrossRef]
18. Harrod, C.; Grey, J.; McCarthy, T.K.; Morrissey, M. Stable isotope analyses provide new insights into ecological plasticity in a mixohaline population of European eel. *Oecologia* **2005**, *144*, 673–683. [CrossRef]
19. Chaouch, H.; Ben Abdallah-Ben Hadj Hamida, O.; Ghorbel, M.; Jarboui, O. Feeding habits of the annular seabream, *Diplodus annularis* (Linnaeus, 1758) (Pisces: Sparidae), in the Gulf of Gabes (Central Mediterranean). *Cahiers de Biologie Marine* **2014**, *55*, 13–19.

20. Vizzini, S.; Mazzola, A. Seasonal variations in the stable carbon and nitrogen isotope ratios ($^{13}C/^{12}C$ and $^{15}N/^{14}N$) of primary producers and consumers in a western Mediterranean coastal lagoon. *Mar. Biol.* **2003**, *142*, 1009–1018. [CrossRef]
21. Persic, A.; Roche, H.; Ramade, F. Stable carbon and nitrogen isotope quantitative structural assessment of dominant species from the Vaccarès Lagoon trophic web (Camargue Biosphere Reserve, France). *Estuar. Coast. Shelf Sci.* **2004**, *60*, 261–272. [CrossRef]
22. Mariani, S.; Maccaroni, A.; Massa, F.; Rampacci, M.; Tancioni, L. Lack of consistency between the trophic interrelationships of five sparid species in two adjacent central Mediterranean coastal lagoons. *J. Fish Biol.* **2002**, *61*, 138–147. [CrossRef]
23. Pita, C.; Gamito, S.; Erzini, K. Feeding habits of the gilthead seabream (*Sparus aurata*) from the Ria Formosa (southern Portugal) as compared to the black seabream (*Spondyliosoma cantharus*) and the annular seabream (*Diplodus annularis*). *J. Appl. Ichthyol.* **2002**, *18*, 81–86. [CrossRef]
24. Bouchereau, J.L.; Marques, C.; Pereira, P.; Guélorget, O.; Lourié, S.M.; Vergne, Y. Feeding behaviour of *Anguilla anguilla* and trophic resources in the Ingril Lagoon (Mediterranean, France). *Cahiers de Biologie Marine* **2009**, *50*, 319.
25. Bouchereau, J.L.; Marques, C.; Pereira, P.; Guélorget, O.; Vergne, Y. Food of the European eel *Anguilla anguilla* in the Mauguio lagoon (Mediterranean, France). *Acta Adriat.* **2009**, *50*, 159–170.
26. Pasquaud, S.; Elie, P.; Jeantet, C.; Billy, I.; Martinez, P.; Girardin, M. A preliminary investigation of the fish food web in the Gironde estuary, France, using dietary and stable isotope analyses. *Estuar. Coast. Shelf Sci.* **2008**, *78*, 267–279. [CrossRef]
27. França, S.; Vasconcelos, R.P.; Tanner, S.; Máguas, C.; Costa, M.J.; Cabral, H.N. Assessing food web dynamics and relative importance of organic matter sources for fish species in two Portuguese estuaries: A stable isotope approach. *Mar. Environ. Res.* **2011**, *72*, 204–215. [CrossRef]
28. Woodward, G.; Hildrew, A.G. Invasion of a stream food web by a new top predator. *J. Anim. Ecol.* **2001**, *70*, 273–288. [CrossRef]
29. Peterson, B.J.; Fry, B. Stable Isotopes in Ecosystem Studies. *Annu. Rev. Ecol. Syst.* **1987**, *18*, 293–320. [CrossRef]
30. Kwak, T.J.; Zedler, J.B. Food web analysis of southern California coastal wetlands using multiple stable isotopes. *Oecologia* **1997**, *110*, 262–277. [CrossRef]
31. Rossi, L.; Sporta Caputi, S.; Calizza, E.; Careddu, G.; Oliverio, M.; Schiaparelli, S.; Costantini, M.L. Antarctic food web architecture under varying dynamics of sea ice cover. *Sci. Rep.* **2019**, *9*, 1–13. [CrossRef]
32. Fry, B. *Stable Isotope Ecology*; Springer: New York, NY, USA, 2006; Volume 521.
33. Fry, B. Coupled N, C and S stable isotope measurements using a dual-column gas chromatography system. *Rapid Commun. Mass Spectrom.* **2007**, *21*, 750–756. [CrossRef]
34. Rossi, L.; Costantini, M.L.; Carlino, P.; di Lascio, A.; Rossi, D. Autochthonous and allochthonous plant contributions to coastal benthic detritus deposits: A dual-stable isotope study in a volcanic lake. *Aquat. Sci.* **2010**, *72*, 227–236. [CrossRef]
35. Careddu, G.; Calizza, E.; Costantini, M.L.; Rossi, L. Isotopic determination of the trophic ecology of a ubiquitous key species—The crab *Liocarcinus depurator* (Brachyura: Portunidae). *Estuar. Coast. Shelf Sci.* **2017**, *191*, 106–114. [CrossRef]
36. Calizza, E.; Careddu, G.; Sporta Caputi, S.; Rossi, L.; Costantini, M.L. Time- and depth wise trophic niche shifts in Antarctic benthos. *PLoS ONE* **2018**, *13*, e0194796. [CrossRef]
37. Post, D.M. Using Stable Isotopes to Estimate Trophic Position: Models, Methods, and Assumptions. *Ecology* **2002**, *83*, 703–718. [CrossRef]
38. Bentivoglio, F.; Calizza, E.; Rossi, D.; Carlino, P.; Careddu, G.; Rossi, L.; Costantini, M.L. Site-scale isotopic variations along a river course help localize drainage basin influence on river food webs. *Hydrobiologia* **2016**, *770*, 257–272. [CrossRef]
39. Dailer, M.L.; Knox, R.S.; Smith, J.E.; Napier, M.; Smith, C.M. Using δ15N values in algal tissue to map locations and potential sources of anthropogenic nutrient inputs on the island of Maui, Hawai'i, USA. *Mar. Pollut. Bull.* **2010**, *60*, 655–671. [CrossRef]

40. Rossi, L.; Calizza, E.; Careddu, G.; Rossi, D.; Orlandi, L.; Jona-Lasinio, G.; Aguzzi, L.; Costantini, M.L. Space-time monitoring of coastal pollution in the Gulf of Gaeta, Italy, using $\delta^{15}N$ values of Ulva lactuca, landscape hydromorphology, and Bayesian Kriging modelling. *Mar. Pollut. Bull.* **2018**, *126*, 479–487. [CrossRef]
41. Orlandi, L.; Calizza, E.; Careddu, G.; Carlino, P.; Costantini, M.L.; Rossi, L. The effects of nitrogen pollutants on the isotopic signal ($\delta^{15}N$) of *Ulva lactuca*: Microcosm experiments. *Mar. Pollut. Bull.* **2017**, *115*, 429–435. [CrossRef]
42. Rosecchi, E. L'alimentation de *Diplodus annularis*, *Diplodus sargus*, *Diplodus vulgaris* et *Sparus aurata* (Pisces, Sparidae) dans le golfe du Lion et les lagunes littorales. *Revue des Travaux de l'Institut des Pêches maritimes* **1987**, *49*, 125–141.
43. Rodriguez-Ruiz, S.; Sànchez-Lizaso, J.L.; Ramos-Esplà, A.A. Feeding of *Diplodus annularis* in *Posidonia oceanica* meadows: Ontogenetic, diel and habitat related dietary shifts. *Bull. Mar. Sci.* **2002**, *71*, 1353–1360.
44. Lazio Region, Piano di Tutela delle Acque Regionale (PTAR) Aggiornamento. Available online: http://www.regione.lazio.it/binary/prl_ambiente/tbl_contenuti/AMB_Piano_tutela_delle_acque_PTAR_aggiornamento.pdf (accessed on 20 March 2020).
45. Cataudella, S.; Crosetti, D.; Massa, F. *Mediterranean Coastal Lagoons: Sustainable Management and Interactions among Aquaculture, Capture Fisheries and the Environment*; General Fisheries Commission for the Mediterranean: Studies and Reviews; FAO: Rome, Italy, 2015; Volume 95.
46. Basset, A.; Galuppo, N.; Sabetta, L. Environmental heterogeneity and benthic macroinvertebrate guilds in Italian lagoons. *Transit. Waters Bull.* **2007**, *1*, 48–63.
47. Cioffi, F.; Gallerano, F. Management strategies for the control of eutrophication processes in Fogliano lagoon (Italy): A long-term analysis using a mathematical model. *Appl. Math. Model.* **2001**, *25*, 385–426. [CrossRef]
48. Jacob, U.; Mintenbeck, K.; Brey, T.; Knust, R.; Beyer, K. Stable isotope food web studies: A case for standardized sample treatment. *Mar. Ecol. Progr.* **2005**, *287*, 251–253. [CrossRef]
49. McCutchan, J.H.; Lewis, W.M.; Kendall, C.; McGrath, C.C. Variation in trophic shift for stable isotope ratios of carbon, nitrogen, and sulfur. *Oikos* **2003**, *102*, 378–390. [CrossRef]
50. Vander Zanden, M.J.; Rasmussen, J.B. A trophic position model of pelagic food webs: Impact on contaminant bioaccumulation in lake trout. *Ecol. Monogr.* **1996**, *66*, 451–477. [CrossRef]
51. Hammer, Ø.; Harper, D.A.T.; Ryan, P.D. *PAST: Paleontological Statistics*; Version 3.0; National History Museum, University of Oslo: Oslo, Norway, 2013.
52. Parnell, A.; Inger, R. *Simmr: A Stable Isotope Mixing Model*; R Package Version 0.3. R. 2016. Available online: https://CRAN.R-project.org/package=simmr (accessed on 20 June 2019).
53. DeNiro, M.J.; Epstein, S. Influence of diet on the distribution of carbon isotopes in animals. *Geochim. Cosmochim. Acta* **1978**, *42*, 495–506. [CrossRef]
54. Minagawa, M.; Wada, E. Stepwise enrichment of ^{15}N along food chains: Further evidence and the relation between $\delta^{15}N$ and animal age. *Geochim. Cosmochim. Acta* **1984**, *48*, 1135–1140. [CrossRef]
55. Bardonnet, A.; Riera, P. Feeding of glass eels (*Anguilla anguilla*) in the course of their estuarine migration: New insights from stable isotope analysis. *Estuar. Coast. Shelf Sci.* **2005**, *63*, 201–209. [CrossRef]
56. Costantini, M.L.; Carlino, P.; Calizza, E.; Careddu, G.; Cicala, D.; Sporta Caputi, S.; Fiorentino, F.; Rossi, L. The role of alien fish (the centrarchid *Micropterus salmoides)* in lake food webs highlighted by stable isotope analysis. *Freshw. Biol.* **2018**, *63*, 1130–1142. [CrossRef]
57. McArthur, R.; Levin, R. The limiting similarity, convergence, and divergence of coexisting species. *Am. Nat.* **1967**, *101*, 377–385. [CrossRef]
58. Levins, R. *Evolution in Changing Environments: Some Theoretical Explorations (No. 2)*; Princeton University Press: Princeton, NJ, USA, 1968.
59. Pianka, E.R. Sympatry of Desert Lizards (Ctenotus) in Western Australia. *Ecology* **1969**, *50*, 1012–1030. [CrossRef]
60. Chesson, J. Measuring preference in selective predation. *Ecology* **1978**, *59*, 211–215. [CrossRef]
61. Costantini, M.L.; Rossi, L. Laboratory study of the grass shrimp feeding preferences. *Hydrobiologia* **2010**, *443*, 129–136. [CrossRef]
62. Layman, C.A.; Arrington, D.A.; Montaña, C.G.; Post, D.M. Can Stable Isotope Ratios Provide for Community-Wide Measures of Trophic Structure? *Ecology* **2007**, *88*, 42–48. [CrossRef]

63. Bearhop, S.; Adams, C.E.; Waldron, S.; Fuller, R.A.; Macleod, H. Determining trophic niche width: A novel approach using stable isotope analysis. *J. Anim. Ecol.* **2004**, *73*, 1007–1012. [CrossRef]
64. Jackson, A.L.; Inger, R.; Parnell, A.C.; Bearhop, S. Comparing isotopic niche widths among and within communities: SIBER—Stable Isotope Bayesian Ellipses in R. *J. Ecol.* **2011**, 595–602. [CrossRef]
65. Parnell, A.C.; Inger, R.; Bearhop, S.; Jackson, A.L. Source partitioning using stable isotopes: Coping with too much variation. *PLoS ONE* **2010**, *5*, e9672. [CrossRef]
66. Brey, T.; Rumohr, H.; Ankar, S. Energy content of macrobenthic invertebrates: General conversion factors from weight to energy. *Mar. Ecol. Prog. Ser.* **1988**, *117*, 271–278. [CrossRef]
67. Neuhoff, H.G. Influence of temperature and salinity on food conversion and growth of different Nereis species (Polychaeta, Annelida). *Mar. Ecol. Prog.* **1979**, *1*, e262. [CrossRef]
68. Ryan, P.A. Seasonal and size-related changes in the food of the short-finned eel, *Anguilla australis* in Lake Ellesmere, Canterbury, New Zealand. *Environ. Biol. Fishes* **1986**, *15*, 47–58. [CrossRef]
69. Whittaker, R.H.; Levin, S.A. *Niche: Theory and application*; Stroudsbourg, PA: Dowden, Hutchison and Ross; Distributed by Halsted Press: New York, NY, USA, 1975.
70. Calizza, E.; Costantini, M.L.; Careddu, G.; Rossi, L. Effect of habitat degradation on competition, carrying capacity, and species assemblage stability. *Ecol. Evol.* **2017**, *7*, 5784–5796. [CrossRef] [PubMed]
71. McClelland, J.W.; Valiela, I.; Michener, R.H. Nitrogen-stable isotope signatures in estuarine food webs: A record of increasing urbanization in coastal watersheds. *Limnol. Oceanogr.* **1997**, *42*, 930–937. [CrossRef]
72. Di Lascio, A.; Rossi, L.; Carlino, P.; Calizza, E.; Rossi, D.; Costantini, M.L. Stable isotope variation in macroinvertebrates indicates anthropogenic disturbance along an urban stretch of the river Tiber (Rome, Italy). *Ecol. Indic.* **2013**, *28*, 107–114. [CrossRef]
73. Vizzini, S.; Mazzola, A. The effects of anthropogenic organic matter inputs on stable carbon and nitrogen isotopes in organisms from different trophic levels in a southern Mediterranean coastal area. *Sci. Total Environ.* **2006**, *368*, 723–731. [CrossRef]
74. Calizza, E.; Aktan, Y.; Costantini, M.L.; Rossi, L. Stable isotope variations in benthic primary producers along the Bosphorus (Turkey): A preliminary study. *Mar. Pollut. Bull.* **2015**, *97*, 535–538. [CrossRef]
75. Vizzini, S.; Mazzola, A. The fate of organic matter sources in coastal environments: A comparison of three Mediterranean lagoons. *Hydrobiologia* **2008**, *611*, 67–79. [CrossRef]
76. Tapia González, F.U.; Herrera-Silveira, J.A.; Aguirre-Macedo, M.L. Water quality variability and eutrophic trends in karstic tropical coastal lagoons of the Yucatán Peninsula. *Estuar. Coast. Shelf Sci.* **2008**, *76*, 418–430. [CrossRef]
77. Tagliapietra, D.; Pavan, M.; Wagner, C. Macrobenthic Community Changes Related to Eutrophication in Palude della Rosa (Venetian Lagoon, Italy). *Estuar. Coast. Shelf Sci.* **1998**, *47*, 217–226. [CrossRef]
78. Mehner, T.; Arlinghaus, R.; Berg, S.; Dörner, H.; Jacobsen, L.; Kasprzak, P.; Koschel, R.; Schulze, T.; Skov, C.; Wolter, C.; et al. How to link biomanipulation and sustainable fisheries management: A step-by-step guideline for lakes of the European temperate zone. *Fish. Manag. Ecol.* **2004**, *11*, 261–275. [CrossRef]
79. Cicala, D.; Calizza, E.; Careddu, G.; Fiorentino, F.; Sporta Caputi, S.; Rossi, L.; Costantini, M.L. Spatial variation in the feeding strategies of Mediterranean fish: Flatfish and mullet in the Gulf of Gaeta (Italy). *Aquat. Ecol.* **2019**, *53*, 529–541. [CrossRef]
80. Dörner, H.; Skov, C.; Berg, S.; Schulze, T.; Beare, D.J.; Van der Velde, G. Piscivory and trophic position of *Anguilla anguilla* in two lakes: Importance of macrozoobenthos density. *J. Fish Biol.* **2009**, *74*, 2115–2131. [CrossRef]
81. Deudero, S.; Pinnegar, J.K.; Polunin, N.V.C.; Morey, G.; Morales-Nin, B. Spatial variation and ontogenic shifts in the isotopic composition of Mediterranean littoral fishes. *Mar. Biol.* **2004**, *145*, 971–981. [CrossRef]
82. Jennings, S.; Barnes, C.; Sweeting, C.J.; Polunin, N.V. Application of nitrogen stable isotope analysis in size-based marine food web and macroecological research. *Rapid Commun. Mass Spectrom.* **2008**, *22*, 1673–1680. [CrossRef]
83. Galvan, D.E.; Sweeting, C.J.; Reid, W.D.K. Power of stable isotope techniques to detect size-based feeding in marine fishes. *Mar. Ecol. Prog. Ser.* **2010**, *407*, 271–278. [CrossRef]
84. Lammens, E.H.R.R.; Nie, H.W.; de Vijverberg, J.; Densen, W.L.T. Resource Partitioning and Niche Shifts of Bream (*Abramis brama*) and Eel (*Anguilla anguilla*) Mediated by Predation of Smelt (*Osmerus eperlanus*) on *Daphnia hyalina*. *Can. J. Fish. Aquat. Sci.* **1985**, *42*, 1342–1351. [CrossRef]

85. De Nie, H.W. Food, feeding periodicity and consumption of the eel *Anguilla anguilla* (L.) in the shallow eutrophic Tjeukemeer (The Netherlands). *Archiv für Hydrobiologie* **1987**, *109*, 421–443.
86. Polis, G.A.; Strong, D.R. Food Web Complexity and Community Dynamics. *Am. Nat.* **1996**, *147*, 813–846. [CrossRef]
87. Layman, C.A.; Araujo, M.S.; Boucek, R.; Hammerschlag-Peyer, C.M.; Harrison, E.; Jud, Z.R.; Matich, P.; Rosenblatt, A.E.; Vaudo, J.J.; Yeager, L.A.; et al. Applying stable isotopes to examine food-web structure: An overview of analytical tools. *Biol. Rev.* **2012**, *87*, 545–562. [CrossRef]
88. Calizza, E.; Costantini, M.L.; Carlino, P.; Bentivoglio, F.; Orlandi, L.; Rossi, L. *Posidonia oceanica* habitat loss and changes in litter-associated biodiversity organization: A stable isotope-based preliminary study. *Estuar. Coast. Shelf Sci.* **2013**, *135*, 137–145. [CrossRef]
89. Calizza, E.; Costantini, M.L.; Rossi, D.; Carlino, P.; Rossi, L. Effects of disturbance on an urban river food web. *Freshw. Biol.* **2012**, *57*, 2613–2628. [CrossRef]

Article

Pretreatment Method for DNA Barcoding to Analyze Gut Contents of Rotifers

Hye-Ji Oh [1], Paul Henning Krogh [2], Hyun-Gi Jeong [3], Gea-Jae Joo [4], Ihn-Sil Kwak [5,6], Sun-Jin Hwang [1], Jeong-Soo Gim [4], Kwang-Hyeon Chang [1,*] and Hyunbin Jo [6,*]

[1] Department of Environmental Science and Engineering, Kyung Hee University, Yongin 17104, Korea; ohg2090@naver.com (H.-J.O.); sjhwang@khu.ac.kr (S.-J.H.)
[2] Department of Bioscience, Aarhus University, Vejlsøvej 25, 8600 Silkeborg, Denmark; phk@bios.au.dk
[3] Nakdong River Environment Research Center, National Institute of Environmental Research, Goryeong 40103; jhgpl@korea.kr
[4] Department of Integrated Biological Science, Pusan National University, Busan 46241, Korea; gjjoo@pusan.ac.kr (G.-J.J.); kjs1@pusan.ac.kr (J.-S.G.)
[5] Faculty of Marine Technology, Chonnam National University, Yeosu 59626, Korea; inkwak@hotmail.com
[6] Fisheries Science Institute, Chonnam National University, Yeosu 59626, Korea
* Correspondence: chang38@khu.ac.kr (K.-H.C.); prozeva@jnu.ac.kr (H.J.); Tel.: +82-10-8620-4184 (K.-H.C.); +82-10-8807-7290 (H.J.)

Received: 29 December 2019; Accepted: 26 January 2020; Published: 5 February 2020

Abstract: We designed an experiment to analyze the gut content of Rotifera based on DNA barcoding and tested it on *Asplanchna* sp. in order to ensure that the DNA extracted from the rotifer species is from the food sources within the gut. We selected ethanol fixation (60%) to minimize the inflow effects of treated chemicals, and commercial bleach (the final concentration of 2.5%, for 210 s) to eliminate the extracellular DNA without damage to the lorica. Rotifers have different lorica structures and thicknesses. Therefore, we chose a pretreatment method based on *Asplanchna* sp., which is known to have weak durability. When we used the determined method on a reservoir water sample, we confirmed that the DNA fragments of Chlorophyceae, Diatomea, Cyanobacteria, and Ciliophora were removed. Given this result, Diatomea and cyanobacteria, detected from *Asplanchna*, can be considered as gut contents. However, bacteria were not removed by bleach, thus there was still insufficient information. Since the results of applying commercial bleach to rotifer species confirmed that pretreatment worked effectively for some species of rotifers food sources, in further studies, it is believed to be applicable to the gut contents analysis of more diverse rotifers species and better DNA analysis techniques by supplementing more rigorous limitations.

Keywords: gut content of Rotifera; eliminate the extracellular DNA; commercial bleach; pretreatment

---

1. Introduction

It is important to understand the role and function of interactions in the microbial food web of aquatic ecosystems. The key biological interaction in the aquatic microbial food web is matter cycling mediated by predation, and predation often works as a regulating factor for energy pathways, as well as determining species composition [1]. In particular, rotifers are critical components linking microorganisms with larger predatory organisms such as crustaceans and fish within the grazing food chain: bacteria, heterotrophic nano-flagellates, rotifers/copepods/cladocerans, larval fish, mature fish [2,3]. Consequently, they function as a channel for the flux of organic matter within diverse organism assemblages organized in an intermediate position between the two different food webs, and transfer nutrients and energy from the microbial loop to higher trophic levels [4–6]. In addition, as the problem of eutrophication increases in aquatic ecosystems, the abundance of macrozooplankton decreases and consequently the contribution of rotifers in energy flow of aquatic food web becomes greater [2]. As a result, rotifer-focused biological interactions, especially rotifer feeding behaviors in microbial food web, are receiving a great attention to understand not only the interrelated biological relationships but also the structure and function in aquatic food webs [7].

However, the comprehensive understanding of rotifers feeding characteristics has not been well-elucidated in comparison to their importance, because most previous studies were conducted at the lab-scale with limited environmental conditions over a short time period, limited to common and dominant species as the tested species, and therefore have not been verified in the field [8–12]. These limitations were attributed to the absence of adequate analytical methodologies applicable to field sites due to difficulties in culturing, handling and identification of both prey and predator (rotifers) which have small sizes (usually rotifers body size$\leq$ 1000 μm; rotifers prey size spectrum<1–20 μm) [13]. In order to overcome the methodological limitations for the analysis of rotifer feeding behaviors, the introduction and application of appropriate techniques are required.

In recent years, genomic technologies have developed rapidly and been applied to ecological research. DNA barcoding techniques have increased the reliability of identifying specific taxonomic groups of organisms at both species and genus level [14], and environmental DNA techniques have enabled the detection of elusive species in various environments [15,16]. Genomic approaches have also been used to understand trophic ecology, particularly biological interaction, for both aquatic habitat environments and food webs by collecting information from food material found in gut contents and the excrement of various organisms and this helps to overcome the existing limitations of food source analyses, which were usually based on visual analysis [17–20].

So far, however, the microscopic and DNA identification of food remains in the gut contents have been limited to large-size organisms such as fish and benthic macroinvertebrates as gut contents extraction is difficult to perform. In the case of zooplankton, crustaceans, with relatively large body size (usually larger than 1 mm) and a hard exoskeleton structure, such as a carapace, which covers the digestive organs, have been the main target for food source analysis. Their morphological characteristics allow physical and chemical treatments, as well as dissection to extract gut contents, avoiding DNA fragments from microorganisms attached to their bodies and DNA from the predator itself. In practice, diets analyses of copepods (small crustaceans) using the DNA-based methods were conducted in both freshwater and ocean ecosystems [21,22]. On the other hand, since small rotifers (usually < 0.5 mm) are relatively soft-bodied, it is difficult to apply similar physical and chemical treatments as for other zooplankton, and there are no proper methodologies and sufficient information of rotifer food sources as results [23]. For a wide range of applications of DNA technology in food source identification, it is necessary to develop a method for separating gut content items from an object by minimizing other possible DNA contaminants, no matter how small the target size is.

For applications of DNA technology to the identification of rotifers food sources, the most critical part of methodology is to distinguish the DNA in the rotifers gut contents from contamination sources that can be attached to the outside of the rotifers lorica and exist in the sample water (so-called ‘extracellular DNA’). Since the detection of extracellular DNAs can cause confusion in the interpretation of the results of the rotifers gut contents analysis, treatment for eliminating them (so-called ‘pretreatment’) is necessary to obtain the more accurate results of rotifers gut contents analysis. However, unlike crustacean zooplankton, which have a solid carapace, the rotifer body is covered by a lorica, which is relatively softer than the carapace. In addition, lorica hardness is differs by species [24].

In this study, we focused on the establishment of an appropriate method for detecting the DNA of gut contents, which is applicable for soft-bodied rotifers. In this analysis procedure, it is important to eliminate the cells and DNA fragments of microorganisms attached to rotifers in order to extract and analyze only those food sources included in the gut to eliminate extra-cellular DNA contamination. Therefore, we selected chemicals for eliminating different types of extracellular DNA and tested their effects on the lorica of rotifers under different concentration treatments to find the most effective concentration and time for both the preservation of the rotifers and removal of different types of extracellular DNA. Following this, we tested the applicability of gut content analysis to rotifers using DNA technology by verifying whether the DNA fragments of rotifer food sources were eliminated or not when the prescribed treatment method was used.

2. Materials and Methods

2.1. Selection of Chemicals for Eliminating Extracellular DNAs

In order to select the appropriate chemicals to remove the extracellular DNA fragments in detecting DNA of rotifers gut contents, we reviewed the different treatment methods and their procedures found in the literature (Table 1, Table A1).

Table 1. Summary of previous treatment processes for decontamination in DNA analyses.

Target	Treatments and Conditions	Ref.
Contaminated DNA for extinguishing the template activity	Incubation of DNA with a psoralen, 8-methoxypsoralen in the dark for 30 min to overnight and subsequent irradiation of UV (365 nm) for 1 hr.	[25]
Bone for removal of contamination	Washed in sterile distilled water, followed by 10% bleach *	[26]
Teeth for destroying any contaminating DNA on the surface	Soaked in hydrogen peroxide (3–30%) for 10–30 min, rinsed with distilled water, rinsed thoroughly with 10% bleach *, rinsed again with distilled water and UV irradiated for 10 min	
Teeth and cortical bone pieces to prevent extraneous contaminations (dirt, carbonate deposits, acid residues)	Soaked for 10 min in 15 % HCL, for 10 min in 70 % ethanol and rinsed in sterile double-distilled water for 30 min	[27]
Tooth for prevention of contamination	Soaked for ~ 10 min in 10% bleach * and then rinsed with 70 % ethanol	[28]
Skeletal material (e.g., powdered bone) for reducing DNA contamination	- Immersing in 20 % bleach * for 2 min followed by extensive ddH_2O washing - 2 days treatment with 0.5 M EDTA at 55 °C	[29]
Bone fragments for elimination of any minor surface contamination	10 min on each side with UV light (254 nm) and soaked for approximately 5 min in a 5 % bleach solution* (in some cases)	[30]
Teeth for removal of dirt and other contaminants	Treated with 30 % acetic acid, rinsed with ultrapure water, immersed for 10 min in 10 % sodium hypochlorite* and exposed to UV light (254 nm) for at least 10 min on each side	[31]
Environmental samples for removal of free extracellular DNA	Added Ethidium or Propidium Monoazied (EMA or PMA) following a conventional procedure in accordance with the manufacturer's protocol (PowerSoil DNA extraction kit (MoBio Lab, Inc.)	[32]
External DNA contamination of arthropod gut-content	40 min of end-over-end rotation in 2.5 % commercial bleach *	[33]

* Underlined treatment and procedure: the case of the application of bleach for decontamination

Most previous published treatment methods were not suitable for the selective elimination of external DNA, which is required for gut content analysis of rotifers, due to their soft lorica. Therefore, we selected ethanol and commercial bleach to remove extracellular DNA from the rotifers, while maintaining the internal gut content DNA composition inside the rotifers. Ethanol (Ethyl alcohol; CAS No. 64-17-5) was used for preservation and sterilization of the raw water sample. Accordingly, the raw water sample was fixed with ethanol to a final concentration of 60% (typical concentrations for disinfection and sterilization: 60–95% [34]). Preserving rotifers with ethanol would limit the exposure to damage of the rotifer lorica by the action of commercial bleach, which has previously been used for DNA elimination and extraction of gut contents in zooplankton, only externally [33,35]. Yuhan-Clorox

(Yuhanrox regular) (Yuhan Co, Ltd., Korea) composed of 4%–6% NaClO (CAS No. 7681-52-9) and 0.1%–0.5 % NaOH (CAS No. 1310-73-2) was used for the chemical wash treatment.

2.2. Responses of Rotifers Lorica to Bleach

To find a suitable treatment time and concentration of commercial bleach for extracellular DNA removal without affecting rotifer lorica and gut contents, we measured the response time of different rotifer species, which were collected from a water reservoir, to different exposure concentrations. We minimized contamination by separating each sample, using bleach sterilized gloves, and instruments sterilized by autoclave and ethanol. As testing the response time and range of commercial bleach concentrations requires multiple individuals of each rotifer species, we targeted large species in which at least three individuals can be gathered by sorting, *Brachionus forficula*, *Keratella* sp., *Trichocerca* sp., *Polyarthra* sp., and *Asplanchna* sp. having variable lorica characteristics; from species having soft lorica (e.g., *Asplanchna* sp.) to hard lorica (e.g., *Keratella* sp. and *Polyarthra* sp.) [36]. Each rotifer species was treated using commercial bleach with final concentrations of 20%, 10%, 5%, and 2.5%. We measured the time until lorica disintegration of the three individuals per rotifer species through microscopic inspection (OLYMPUS CKX41). These time results were used as baselines to determine the concentration and time of removing extracellular DNA without damaging the rotifer individuals (Figure 1).

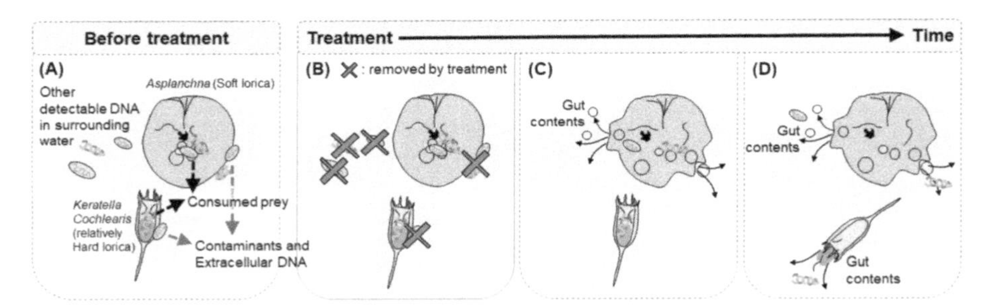

Figure 1. Diagram of necessity of proper pretreatment process in rotifers gut contents analysis; (**A**) Without treatment for removing contaminants and other detectable DNA including extracellular DNA, there is possibility to be confused that detected DNA is from the rotifers gut contents or not, (**B**) Through proper treatment, contaminants and other detectable DNA can be removed without damaging the rotifer individuals, (**C**) and (**D**) Rotifer individuals can be damaged and their gut contents can be overflowed by excessive treatment according to their lorica characteristics.

2.3. Application and Effectiveness Verification of Set Treatment Concentration and Time

We collected water samples from a eutrophic reservoir (Shin-gal reservoir, Korea; N 37.241536, E 127.0929190) in fall (4th November 2018). Rotifers dominated the zooplankton community of the reservoir during this season. We repeatedly filtered 10 L of surface water (n ≒ 20) into a 60 μm mesh sized zooplankton net and obtained a 1 L filtered water sample. From the collected sample, all organisms were removed by hand using a microscope (OLYMPUS CKX41) and 0.5 mL of subsample was extracted to micro-tubes. For every sample taken, we made a negative control to prevent cross contamination.

For the application and effective verification of a set treatment concentration and time on the DNA fragments of rotifer food sources (Chlorophyceae, Diatomea, Cyanobacteria, Bacteria, Ciliophora, and Heterotrophic nanoflagellates) [9], we compared both treated and non-treated samples with 0.5 mL filtered water samples. In the case of treated sample, after set treatment time, we poured and filtered the sample immediately by washing with distilled water to prevent further effects. Non-treated sample was also filtered in order to proceed with the same DNA extraction process as the treated sample.

To confirm detection of DNA in the rotifers gut without extracellular DNA using the suggested treatment method, we collected a rotifer species, *Asplanchna* sp., from reservoir water, and applied this method to the treatment. We sorted *Asplanchna* specimens from the water sample and transferred them to distilled water several times until pure rotifer individuals were collected without other visible particles, particularly phytoplankton cells. We then checked for removal of particles under the microscope and selected clean individuals without attached particle or microorganisms (one individual per a sample, 3 replicates). As with water samples, rotifers samples were filtered after pretreatment for extraction of their DNA. The 47mm diameter cellulose nitrate filters with a pore size of 0.45μm pore size (NC 45 ST, Whatman^{TM}) were used to filter DNA fragments [37].

2.4. DNA Analysis Precedure

The DNA was isolated from the filter paper using DNeasy Blood & Tissue kit (Qiagen, Hilden, Germany) according to the manufacturer's instructions on a clean bench (Supplementary Materials). In order to reduce potential contamination during DNA extraction and the amplification process, we minimized contamination sources by separating each sample, using bleach sterilized gloves, and instruments sterilized by autoclave and ethanol.

A polymerase chain reaction (PCR) amplification was performed using AccuPower Hot start PCR PreMix (Bioneer, Korea) with genomic DNA and primers in a final volume of 20 μL. Primers were selected for specific detection of each potential prey community for rotifers. Phytoplankton and the components of the microbial food web (bacteria and protozoa) were considered as potential food sources for rotifers [6]. At this time, since we could not find a suitable primer to detect only heterotrophic nanoflagellates (HNF) specifically, we applied a universal primer for eukaryotes to HNF [38,39]. PCR conditions for each primer in a thermal cycler (Bio-rad, California, USA) were summarized in Table 2. PCR products were separated using 1.5% agarose gels (AccuPrep® PCR/Gel DNA Purification Kit (50 reactions) [K-3038]), and the appeared band from PCR products was extracted and sequenced in both directions by capillary sequencing at Bioneer Co. (Daejeon, Korea). Additionally, cloning was carried out using the pGEM-T easy vector (Promega, Madison, USA) to confirm direct sequencing results. Cloned plasmid DNA was isolated according to the alkaline-lysis method using Labopass Plasmid Miniprep kit (Cosmogenetech). Individually isolated plasmid DNA was then digested using the restriction enzyme EcoRI to confirm insertion. Positive clones for each sample were analyzed to species-specific sequences with SP6 primers using an automated 3730 DNA analyzer (Applied Biosystems, Foster City, USA).

Sequence alignment was performed using Clustal W 2.0 [40]. A BLASTn [41] search was performed to identify sequences with the best hits. In the GenBank nucleotide collection database, the organisms, which were included in the database search, were optimized for highly similar sequences by BLASTn, and selected by high identity (%).

Table 2. PCR primers used for detecting rotifer food sources.

Food source	Primer	Sequence (5′-3′)	Base Pair	Ref.
Chlorophyceae (18s rRNA)	ChloroF	TGG CCT ATC TTG TTG GTC TGT	473 bp	[42]
	ChloroR	GAA TCA ACC TGA CAA GGC AAC		
	94 °C, 3 min -> 35cycles [94 °C, 1 min -> 55 °C **, 1 min -> 72 °C, 1 min] -> 72 °C, 10 min			
Diatomea (18s rRNA; V4)	M13F-D512for	TGT AAA ACG GCC AGT ATT CCA GCT CCA ATA GCG	390–410 bp	[43]
	M13R-D978rev	CAG GAA ACA GCT ATG ACG ACT ACG ATG GTA TCT AAT C		
	94 °C, 2 min ->5cycles [94 °C, 45s -> 53°C **, 45s -> 72 °C, 1 min] ->35cycles [94 °C, 45 s -> 51°C **, 45s -> 72 °C, 1 min] -> 72 °C, 10 min			

Table 2. *Cont.*

Food source	Primer	Sequence (5′-3′)	Base Pair	Ref.
Cyanobacteria (16S rRNA; ITS *)	16S27F	AGA GTT TGA TCC TGG CTC AG	422 bp	[44]
	23S30R	CTT CGC CTC TGT GTG CCT AGG T		
		94 °C, 5min -> 10cycles [94 °C, 45s -> 57 °C **, 45s -> 68 °C, 2 min] -> 25cycles [92 °C, 45s -> 54 °C **, 45s -> 68 °C, 2 min] -> 68 °C, 7 min		
Bacteria (16S rDNA; nearly full-length)	Forward	GAG TTG GAT CCT GGC TCA G	About 2000 bp	[45,46]
	Reverse	AAG GAG GGG ATC CAG CC		
		95 °C, 3 min -> 35cycles [94 °C, 1 min -> 60 °C **, 1 min -> 72 °C, 2 min] -> 72 °C, 3 min		
Ciliophora (18S rRNA)	Cil F	TGG TAG TGT ATT GGA CWA CCA	600–670 bp	[47]
	Cil R- 1	TCT GAT CGT CTT TGA TCC CTT		
	Cil R- 2	TCT RAT CGT CTT TGA TCC CCT A		
	Cil R- 3	TCT GAT TGT CTT TGA TCC CCT		
		95 °C, 5 min -> 35cylces [94 °C, 45 s -> 55 °C **, 1min -> 72 °C, 1 min] -> 72°C, 10min		
Heterotrophic nanoflagellates (18s rRNA)	EukA ***	AAC CTG GTT GAT CCT GCC AGT	800–900 bp	[38,39]
	EukB ***	TGA TCC TTC TGC AGG TTC ACC TAC		
		95 °C, 2 min -> 35cycles [95°C, 30s -> 55 °C **, 30s -> 72 °C, 2 min] -> 72 °C, 7 min		

* Internal transcribed spacer; ** Underlined temperature: annealing temperature of each primer; *** EukA and EukB are universal primer for eukaryote.

3. Results

3.1. Responses of Rotifers Lorica to Commercial Bleach Treatment

After treatment with commercial bleach at final concentrations of 20%, 10%, 5%, and 2.5%, the time before the loss of each rotifer's contents by lorica disintegration (five rotifers species tested, n=3) was measured. Every tested rotifer species, *Brachionus forficula*, *Keratella* sp., *Trichocerca* sp., *Polyarthra* sp., and *Asplanchna* sp. tended to have shorter times for tolerating treatment as the final concentration of commercial bleach increased. In particular, *Asplanchna* sp. having the weakest lorica showed the shortest time among tested rotifers regardless of treatment concentration. The lorica of *Asplanchna* disintegrated between 35 s and 240 s following exposure to different treatment solutions of various concentration, and its body contents including the gut contents were released from the body. When *Asplanchna* was treated with 2.5% diluted commercial bleach, although it was observed to withstand up to 300 s of exposure, its lorica began to suffer disintegration after 240 s. Therefore, for preservation of its gut contents, the treatment time should be considered as less than 240 s. Other tested rotifer species also showed different duration times for lorica survival against treatment solution depending on its concentration. However, regardless of their lorica thickness and structure, they showed a range of endurance time from 300 to 450 s at 2.5% of commercial bleach (Table 3, Figure 2).

Table 3. The responses of rotifers lorica to commercial bleach treatment; the minimum time (s) before the loss of the rotifers contents by lorica disintegration of each rotifers species.

Concentration (Final) and Duration Time	Rotifer Species				
	Brachionus Forficula	***Keratella* sp.**	***Trichocerca* sp.**	***Polyarthra* sp.**	***Asplanchna* sp.**
20%	60 s	90 s	120 s	60 s	35 s
10%	210 s	180 s	150 s	90 s	45 s
5%	300 s	240 s	300 s	210 s	120 s
2.5%	450 s	300 s	450 s	300 s	240 s

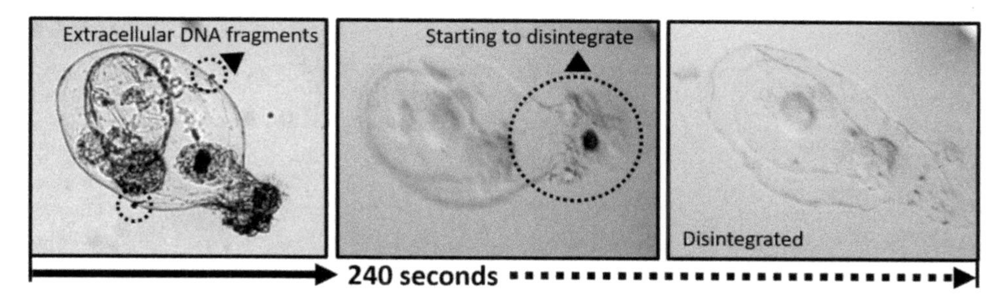

Figure 2. An example of a disintegration process; response of *Asplanchna* sp. lorica to commercial bleach treatment (2.5 %).

To prevent loss of gut contents during pretreatment for extracellular DNA by disintegration of rotifers lorica, we should establish the conditions (concentration and time of chemical) under which extracellular DNA can be removed, and keep the rotifer lorica undamaged. Since high concentration treatments allowed very limited time available for affecting the elimination of extracellular DNA, we decided to use the lowest concentration of commercial bleach for the longest time on samples in order to minimize damage while maximizing external DNA removal. Therefore, based on the response time of *Asplanchna* sp. lorica to the lowest concentration of commercial bleach treatment and consequent its shortest duration time examined by the experiment, rotifer specimens for extracting gut contents DNA were treated with 2.5% diluted commercial bleach for 210 s (Table 3). We observed each treatment process through a microscope, and confirmed that the gut contents of rotifers were likely to be released from the body when their lorica began to disintegrate. Therefore, we judged that it would be appropriate to use commercial bleach for removal of the extracellular DNA up to 30 s before the time when rotifer loricas begin disintegrating. In addition, to maximize treatment time while minimizing internal effects of the treatment by fixing rotifers, we determined that preservation with 60% ethanol soon after the samples are collected and treatment of 2.5% diluted commercial bleach for 210 s was the most effective pretreatment (Figure 3). We, therefore, selected *Asplanchna* sp. for our further experiment. In addition, *Asplanchna sp.* have a typical omnivore feeding behavior according to Chang et al. (2010) [48]. Therefore, it is an ideal experiment creature for this study.

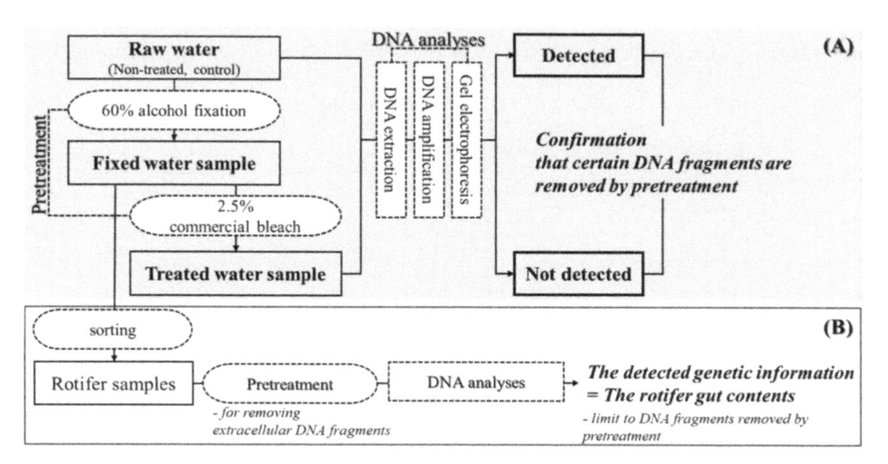

Figure 3. Experimental designs. (**A**) Verifying applicability of pretreatment, (**B**) Application of pretreatment to rotifers.

3.2. Application and Effectiveness Verification of Set Pretreatment Concentration and Time

When the electrophoresis results of raw water from reservoir (non-treated water sample), treated water and treated rotifers samples were compared, they showed different bands in each gel. In the non-treated water sample, the primers used to detect various regions of genetic sequences were all amplified and detected as bands in the electrophoresis gel (Figure 4A–E,N). As a result of identifying the dominant signal information of sequences by the direct capillary sequencing method through BLASTn, all dominant signal identified in non-treated water samples were of Chlorophyceae, Diatomea, Cyanobacteria, Bacteria, and Ciliophora, which are known as common food sources of rotifers (Table 4; Non-treated water sample).

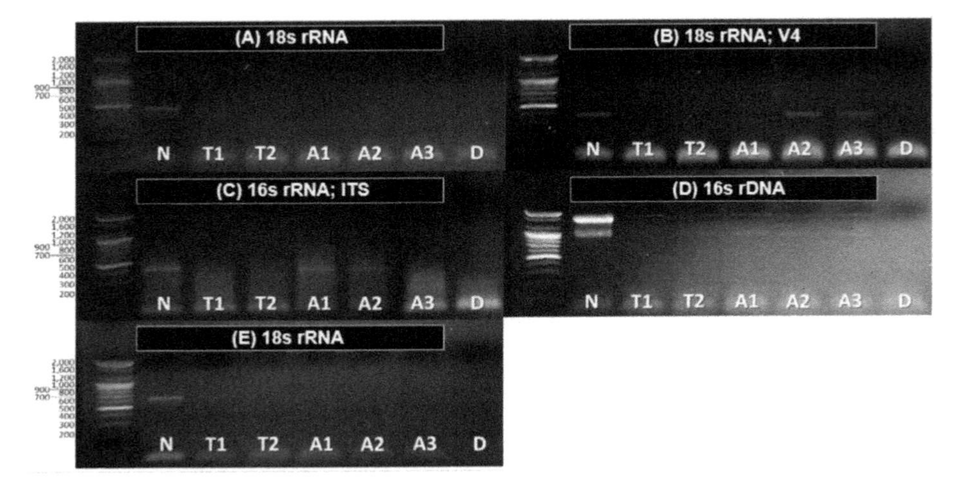

Figure 4. Electrophoresis detection results, (**A**) 18s rRNA for detecting Chlorophyceae; (**B**) 18s rRNA; V4; (**C**) 16s rRNA; ITS; (**D**) 16s rDNA; (**E**) 18s rRNA for detecting Ciliophora and HNF, heterotrophic nanoflagellates; N: non-treated water sample (raw water; control); T: treated water sample by ethanol (60%) and 2.5% commercial bleach solution; A: treated *Asplanchna* samples (n=3); D: distilled water (negative control); first lane of each gel: Ladder using 100-bp molecular marker.

Table 4. The summary of detected dominant signal information based on the direct capillary sequencing and cloning (identity %).

Samples.	18s rRNA	18s rRNA; V4	16S rRNA; ITS	16S rDNA	18S rRNA
Non-treated	*Chlamydomonas nivalis* (99%) *Vitreochlamys nekrassovii* (99%)	*Aulacoseira granulate* (100%) *Aulacoseira ambigua*(99%)	*Chlorophyta sp.* (98%)	*Bacillus* cereus (81%)	*Tintinnidium fluviatile* (90%)
Treated	Not-detected	*Choanoflagellate* (97%) *Meira nashicola* (99%)	Not-detected	*Bacillus thuringiensis* (85%)	Not-detected

* The eukaryote universal primer, which was used to detect HNF (heterotrophic nanoflagellates), has detected Ciliophora (*Tintinnidium* sp.; 88%).

On the other hand, food sources identified in non-treated water sample, except for bacteria were not detected after the treatment of 2.5% diluted commercial bleach for 210 s, indicating their DNA fragments were eliminated by our selected treatment method (Figure 4A–E, T1~T2). Based on these results, we treated same process on *Asplanchna* specimen sorted from the reservoir for verifying if this pretreatment is proper to apply to rotifer species. Most identified species in non-treated water sample were not detected as gut contents in *Asplanchna*. However, Choanoflagellates, fungi species (*Meira* sp.), and bacteria species (*Bacillus* sp.) were detected in some individuals. These species identified in treated *Asplanchna* samples seemed to have been detected by eliminating the signals that were strongly captured from the various DNA fragments that existed before the pretreatment. It means that selected commercial bleach as a pretreatment chemical and specified its concentration and exposure time properly can facilitate the removal of extracellular DNA fragments simultaneously with preserving rotifer body tissue, and consequently this process can be applied for detecting DNA of rotifers gut contents without fear of extracellular DNA contamination (Figure. 4A–E, A1~A3).

Unfortunately, it was difficult to interpret the detected band of bacteria in *Asplanchna* specimen as gut contents, because bacteria were not completely eliminated by the treatment of 2.5% diluted commercial bleach. The sequences from detected bands in the electrophoresis gel let us know that bacteria in non-treated and treated samples are species included in genus *Bacillus* (Table 4; Treated *Asplanchna* sample).

4. Discussion

As a chemical for pretreatment to remove external DNA on the lorica, we selected commercial bleach, which can be used on samples through the proper combination of bleach concentration and exposure duration time (seconds). Commercial bleach has been used mainly to prevent or eliminate contamination in DNA analyses (Table 1), but at the same time, it affects the body tissue of zooplankton, which can lead to the disintegration of rotifer loricas, and thus the release of rotifers' gut contents [33,35]. In this study, it was found that each rotifer species showed different response times when treated with commercial bleach at the same concentration, and the duration time for lorica survival differed by its characteristics (Table 3); the shortest time was for *Asplanchna* sp., having the softest lorica and the longest time for *Trichocerca* sp. and *Brachionus* sp. having lorica that are not easily damaged [49]. In the case of the genus *Keratella*, although its lorica has been suggested as a hard cover, which can be protective against mechanical interference by daphnids and predation by invertebrate predators [36], the loss of inner contents occurred through the mouth parts and not through lorica disintegration following the commercial bleach treatment. Therefore, based on the response time observed for *Asplanchna* sp. showing the shortest response time for disintegration after treatment with 2.5% diluted commercial bleach, we set the pretreatment time to 210 s, as this is an appropriate standard pretreatment method universally applicable to all rotifer taxa.

After applying this pretreatment method to raw water samples from the reservoir, it was confirmed that DNA fragments of rotifer food sources detected in non-treated samples were completely removed; Chlorophyceae, Diatomea, Cyanobacteria, and Ciliophora, except for bacteria (Figure 3, N, T1, T2). Further sequencing analyses indicated that the bacteria detected were mainly *Bacillus* sp. which is known to be tolerant and survive various removal treatments such as disinfection [50] (Table 4). *Bacillus* sp., gram-positive bacterium, has commonly been found in soil and other environments. It has been reported that *Bacillus* plays important roles in the lysis of bloom-forming blue-green alga and the control of their biomass in aquatic ecosystems [51,52]. Therefore, when we applied the pretreatment determined from this study to DNA analysis of gut content of rotifers, we cannot distinguish the source origins of bacteria detected in rotifer species, whether they came from contamination, water samples, or rotifer gut content, like the results of *Asplanchna* (Figure. 4D, A1~3). Since bacteria is one possible main food source for rotifers [53], a suitable pretreatment method for eliminating extracellular bacterial DNA should additionally be developed.

In the results based on DNA analysis, we used each group-specific primer for detection of targeted groups to confirm their presence/absence. As far as we know, there is no information about HNF-specific primer [54], so we applied instead a universal primer set for eukaryotes (Euk-A and Euk-B) which has been used to detect HNF (Table 2). We, therefore, carried out an additional experiment to define the applicability of the HNF primers set to rotifers. As a result, this primer set amplified all possible rotifer species from our study site except for *Asplanchna* sp. (Figure A1). These results provide a proper explanation for why the primer set did not work for all our samples (Figure 4F). The usage of the primer sets that act specifically for each targeted biological community can help in improving detection accuracy for a targeted species group. However, there remain some limitations in verifying the effectiveness of a determined pretreatment on biological communities where specific primers have not yet been developed, such as HNF. In spite of these limitations, the results of the applied pretreatment method to *Asplanchna* sp. showed that specific food sources were detected in the gut content. Choanoflagellate, HNF species, has the habitat selection characteristic of being attached to phytoplankton species, and consequently it is expected that rotifers can eat Choanoflagellate indirectly in the process of eating phytoplankton, or select it as their food source directly [55]. In the case of *Meira nashicola*, which is a kind of yeast-like fungi species, although whether or not *M. nashicola* exists in aquatic ecosystem needs further research, it is considered a valid result of Asplanchna gut contents because parasitic fungus on phytoplankton, such as cyanobacteria, are known to feed on rotifers as alternative food sources [56]. So, when the limitations related to the detection of bacteria and HNF will be resolved, rotifer gut contents can be analyzed by pretreating with alcohol and commercial bleach as we recommend in the present study. Our study used traditional primer sets information; however, Adl et al. (2019) [54] recently revised the classification and nomenclature of Eurkaryotes and recommend some primer sets (rbcL, 18S V4) for Diatomea and Ciliophora. Therefore, we should apply these primer sets according to this new system for further study.

The main goal of the present study was to develop a pretreatment process that eliminated extracellular DNA fragments adhering to the Rotifera lorica and employing DNA barcoding, in order to accurately identify rotifer gut contents, thereby providing a better understanding of rotifer feeding behavior. We devised an experimental design for rotifer gut content analysis on the basis of DNA technology (DNA barcoding) while hypothesizing that feeding behavior (food selectivity) of rotifers with species-specific masticatory apparatus, e.g., the trophi, is dependent on the trophi characteristics. In this process of developing an experimental design, a pretreatment process for removing extracellular DNA as well as the cells attached to the rotifer lorica is essential in isolating accurately the DNA of the food sources remaining within each rotifer gut. Therefore, we selected appropriate chemicals for pretreatment and tried to establish the proper treatment bleach concentration (%) and duration time (seconds) by observing the response time for different types of lorica firstly fixed by 60% diluted ethanol and secondly treated with 2.5% diluted commercial bleach for 210 s. The final pretreatment process was tested on a water sample and a rotifer species (*Asplanchna* sp.) to verify its effectiveness. We conclude that the pretreatment process for rotifer worked effectively in removing extracellular DNA while enabling identification of selected food source taxa of rotifers using DNA barcoding. In this study, single PCR products from group-specific primers and the general eukaryotic primers for HNF were sequenced by the cloning and Sanger method. In forthcoming studies, the taxonomic diversity of the gut content may be analyzed using next-generation sequencing (NGS) while applying improved methods for the decontamination and selection of primers in the controlled experimental environments. The DNA analysis process of rotifer gut contents, especially the pretreatment process, can allow various approaches for DNA analyses for microinvertebrates whose feeding behavior is not sufficiently understood.

Supplementary Materials: The following are available online at http://www.mdpi.com/2076-3417/10/3/1064/s1. DNA sequences: Supplementary_Raw Sequences.FASTA (text format)

Author Contributions: conceptualization, H.-J.O., H.-G.J., and K.-H.C.; methodology, H.J., J.-S.G., K.-H.C. and H.J.; formal analysis, H.J., G.-J.J., S.-J.H. and H.J.; investigation, H.-J.O. and K.-H.C.; writing—original draft preparation, H.-J.O. and K.-H.C.; writing—review and editing, P.H.K., K.-H.C. and H.J.; visualization, H.-J.O., H.-G.J., and K.-H.C.; supervision, K.-H.C. and H.J.; project administration, I.-S.K.; funding acquisition, I.-S.K. and H.J. All authors have read and agreed to the published version of the manuscript.

Funding: This research was funded by the National Research Foundation of Korea, grant number NRF-2018R1A6A1A03024314.

Conflicts of Interest: The authors declare no conflict of interest.

Appendix A

Table A1. Detailed previous treatment procedures in Table 1 (summary of previous treatment processes for decontamination in DNA analyses).

Treatments	Procedure	Ref.
Psoralen + UV irradiation	1. 8-methoxypsoralen of 100 $\mu g \cdot mL^{-1}$ 2. Irradiation with long-wave (365 nm) UV light for 1 h	[25]
Hydrogen peroxide + Bleach * + UV irradiation	1. Soaked in hydrogen peroxide (3–30%) for 10–30 min 2. Rinsed with distilled water 3 *. Rinsed thoroughly with 10% bleach 4. Rinsed with distilled water 5. UV irradiated for 10 min	[26]
Acid wash + Ethanol + UV irradiation	1. Soaked in 15% HCl for 10 min 2. Rinsed with 70% ethanol for 10 min 3. Rinsed in sterile double-distilled water for 30 min 4. UV irradiation (254 nm) for 15 min	[27]
Bleach * + Ethanol	1 *. Soaked in 10% bleach for ~10 min 2. Rinsed with 70% ethanol	[28]
Bleach * + EDTA	1 *. Immersed in 20% bleach for 2 min 2. Rinsed with distilled water 1. 0.5M EDTA at 55°C in a 2-day	[29]
UV irradiation + Bleach *	1. UV irradiation (254 nm) for 10 min 2 *. Soaked for approximately 5 min in a 5% bleach solution	[30]
Acid wash + Bleach * + UV irradiation	1. 30% acetic acid 2. Rinsed with ultrapure water 3 *. Immersed for 10 min in 10% sodium hypochlorite with sporadic shaking 4. Exposed to UV irradiation (254 nm) for at least 10 min	[31]
Ethidium Monoazied (EMA) or Propidium Monoazied (PMA)	1. EMA or PMA added following a conventional procedure in accordance with the manufacturer's protocol (PowerSoil DNA extraction kit (Mo Bio Lab, Inc.)	[32]
Bleach *	1 *. Exposure to 2.5% bleach for 40 min or overnight	[33]

* Underlined treatment and procedure: the case of the application of bleach for decontamination

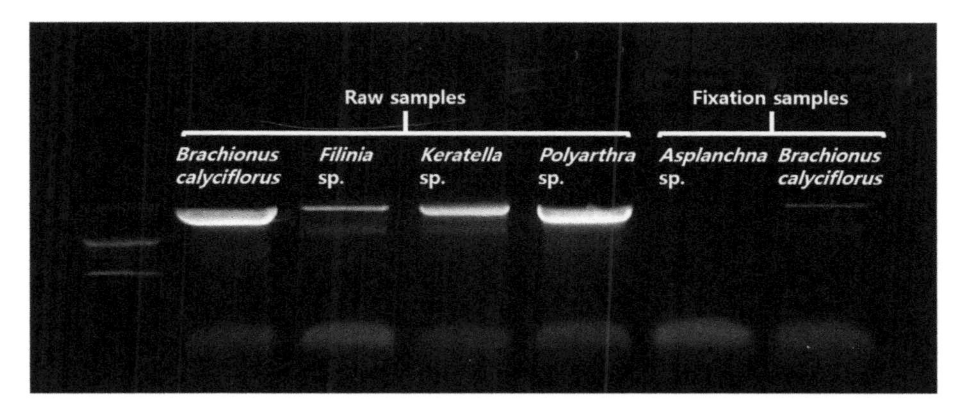

Figure A1. Results of applying Euk universal eukaryotic primers on each rotifer species (Raw samples: collected in the fresh water samples in the Shin-gal reservoir, fixation samples: stored in the laboratory).

References

1. Carrillo, P.; Medina-Sánchez, J.M.; Villar-Argaiz, M.; Delgado-Molina, J.A.; Bullejos, F.J. Complex interactions in microbial food webs: stoichiometric and functional approaches. *Limnetica* **2006**, *25*, 189–204.
2. Wallace, R.L.; Snell, T.W.; Claudia, R.; Thomas, N. *Rotifera vol. 1: Biology, Ecology and Systematics*, 2nd ed.; Backhuys: Leiden, The Nederlands, 2006; p. 94.

3. Pree, B.; Larsen, A.; Egge, J.K.; Simonelli, P.; Madhusoodhanan, R.; Tsagaraki, T.M.; Vage, S.; Erga, S.R.; Bratbak, G.; Thingstad, T.F. Dampened copepod-mediated trophic cascades in a microzooplakton-dominated microbial food web: a mesocosm Study. *Limnol. Oceanogr.* **2017**, *62*, 1031–1044. [CrossRef]
4. Gurav, M.N.; Pejaver, M.K. Survey of rotifers to evaluate the water quality of the river Gadhi and its reservoir. *Ecol. Environ. Conserv.* **2013**, *19*, 417–423.
5. Neto, G.; José, A.; Silva, L.C.D.; Saggio, A.A.; Rocha, O. Zooplankton communities as eutrophication bioindicators in tropical reservoirs. *Biota Neotropica* **2014**, *14*, e20140018.
6. Moore, M.V.; De Stasio, B.T., Jr.; Huizenga, K.N.; Silow, E.A. Trophic coupling of the microbial and the classical food web in Lake Baikal, Siberia. *Freshwater Biol.* **2019**, *64*, 138–151.
7. Oh, H.J.; Jeong, H.G.; Nam, G.S.; Oda, Y.; Dai, W.; Lee, E.H.; Kong, D.; Hwang, S.J.; Chang, K.H. Comparison of taxon-based and trophi-based response patterns of rotifers community to water quality: applicability of the rotifer functional group as an indicator of water quality. *Anim. Cells Syst.* **2017**, *21*, 133–140. [CrossRef]
8. Rothhaupt, K. Differences in particle size-dependent feeding efficiencies of closely related rotifer species. *Limnol. Oceanogr.* **1990**, *35*, 16–23. [CrossRef]
9. Arndt, H. Rotifers as predators on components of the microbial web (bacteria, heterotrophic flagellates, ciliates)—a review. In *Rotifer Symposium VI.*; Gilbert, J.J., Lubzens, E., Miracle, M.R., Eds.; Springer: Dordrecht, The Nederlands, 1993; pp. 231–246.
10. Thouvenot, A.; Debroas, D.; Richardot, M.; Devaux, J. Impact of natural metazooplankton assemblage on planktonic microbial communities in a newly flooded reservoir. *J. Plankton Res.* **1999**, *21*, 179–199. [CrossRef]
11. Mohr, S.; Adrian, R. Reproductive success of the rotifer *Brachionus calyciflorus* feeding on ciliates and flagellates of different trophic modes. *Freshwater Biol.* **2002**, *47*, 1832–1839. [CrossRef]
12. Devetter, M.; Sed'a, J. Rotifer fecundity in relation to components of microbial food web in a eutrophic reservoir. *Hydrobiologia* **2003**, *504*, 167–175. [CrossRef]
13. Pourriot, R. Food and feeding habits of Rotifera. *Arch. Hydrobiol. Beihefte* **1977**, *8*, 243–260.
14. Hebert, P.D.; Cywinska, A.; Ball, S.L.; Dewaard, J.R. Biological identifications through DNA barcodes. *Proc. R. Soc. Lond. Ser. B Biol. Sci.* **2003**, *270*, 313–321. [CrossRef]
15. Barnes, M.A.; Turner, C.R. The ecology of environmental DNA and implications for conservation genetics. *Conserv. Genet.* **2016**, *17*, 1–17. [CrossRef]
16. Kress, W.J.; García-Robledo, C.; Uriarte, M.; Erickson, D.L. DNA barcodes for ecology, evolution, and conservation. *Trends Ecol. Evol.* **2015**, *30*, 25–35. [CrossRef]
17. Symondson, W.O.C. Molecular identification of prey in predator diets. *Mol. Ecol.* **2002**, *11*, 627–641. [CrossRef]
18. Carreon-Martinez, L.; Johnson, T.B.; Ludsin, S.A.; Heath, D.D. Utilization of stomach content DNA to determine diet diversity in piscivorous fishes. *J. Fish Biol.* **2011**, *78*, 1170–1182. [CrossRef] [PubMed]
19. Jo, H.; Gim, J.A.; Jeong, K.S.; Kim, H.S.; Joo, G.J. Application of DNA barcoding for identification of freshwater carnivorous fish diets: Is number of prey items dependent on size class for *Micropterus salmoides*? *Ecol. Evol.* **2014**, *4*, 219–229. [CrossRef]
20. Jo, H.; Ventura, M.; Vidal, N.; Gim, J.S.; Buchaca, T.; Barmuta, L.A.; Jeppesen, E.; Joo, G.J. Discovering hidden biodiversity: the use of complementary monitoring of fish diet based on DNA barcoding in freshwater ecosystems. *Ecol. Evol.* **2016**, *6*, 219–232. [CrossRef]
21. Craig, C.; Kimmerer, W.J.; Cohen, C.S. A DNA-based method for investigating feeding by copepod nauplii. *J. Plankton Res.* **2013**, *36*, 271–275. [CrossRef]
22. Ho, T.W.; Hwang, J.S.; Cheung, M.K.; Kwan, H.S.; Wong, C.K. DNA-based study of the diet of the marine calanoid copepod *Calanus sinicus*. *J. Exp. Mar. Biol. Ecol.* **2017**, *494*, 1–9. [CrossRef]
23. Hochberg, R.; Wallace, R.L.; Walsh, E.J. Soft bodies, hard jaws: an introduction to the symposium, with rotifers as models of jaw diversity. *Integr. Comp. Biol.* **2015**, *55*, 179–192. [CrossRef] [PubMed]
24. Yin, X.; Jin, W.; Zhou, Y.; Wang, P.; Zhao, W. Hidden defensive morphology in rotifers, benefits, costs, and fitness consequences. *Sci. Rep.* **2017**, *7*, 4488.
25. Jinno, Y.; Yoshiura, K.; Niikawa, N. Use of psoralen as extinguisher of contaminated DNA in PCR. *Nucleic Acids Res.* **1990**, *18*, 6739. [CrossRef]
26. Merriwether, D.A.; Rothhammer, F.; Ferrell, R.E. Genetic variation in the New World: ancient teeth, bone, and tissue as sources of DNA. *Experientia* **1994**, *50*, 592–601. [CrossRef] [PubMed]

27. Lalueza, C.; Perez-Perez, A.; Prats, E.; Cornudella, L.; Turbon, D. Lack of founding Amerindian mitochondrial DNA lineages in extinct aborigines from Tierra del Fuego-Patagonia. *Hum. Mol. Gen.* **1997**, *6*, 41–46. [CrossRef] [PubMed]
28. Stone, A.C.; Stoneking, M. mtDNA analysis of a prehistoric Oneota population: implications for the peopling of the New World. *Am. J. Hum. Genet.* **1998**, *62*, 1153–1170. [CrossRef] [PubMed]
29. Kolman, C.J.; Tuross, N. Ancient DNA analysis of human populations. *Am. J. Phys. Anthropol.* **2000**, *111*, 5–23. [CrossRef]
30. Kaestle, F.A.; Smith, D.G. Ancient mitochondrial DNA evidence for prehistoric population movement: The Numic expansion. *Am. J. Phys. Anthropol.* **2001**, *115*, 1–12. [CrossRef]
31. Montiel, R.; Malgosa, A.; Francalacci, P. Authenticating ancient human mitochondrial DNA. *Hum. Biol.* **2001**, *73*, 689–713. [CrossRef]
32. Wagner, A.O.; Malin, C.; Knapp, B.A.; Illmer, P. Removal of free extracellular DNA from environmental samples by ethidium monoazide and propidium monoazide. *Appl. Environ. Microbial.* **2008**, *74*, 2537–2539. [CrossRef]
33. Greenstone, M.H.; Weber, D.C.; Coudron, T.A.; Payton, M.E.; Hu, J.S. Removing external DNA contamination from arthropod predators destined for molecular gut-content analysis. *Mol. Ecol. Resour.* **2012**, *12*, 464–469. [CrossRef]
34. Rutala, W.A.; Weber, D.J. Disinfection, sterilization, and antisepsis: An overview. *Am. J. Infect. Control.* **2016**, *44*, e1–e6. [CrossRef] [PubMed]
35. Laspoumaderes, C.; Modenutti, B.; Balseiro, E. Herbivory versus omnivory: linking homeostasis and elemental imbalance in copepod development. *J. Plankton Res.* **2010**, *32*, 1573–1582. [CrossRef]
36. Gilbert, J.J.; Williamson, C.E. Predator-prey behavior and its effect on rotifer survival in associations of *Mesocyclops edax, Asplanchna girodi, Polyarthra vulgaris,* and *Keratella cochlearis*. *Oecologia* **1978**, *37*, 13–22. [CrossRef] [PubMed]
37. Pilliod, D.S.; Goldberg, C.S.; Arkle, R.S.; Waits, L.P. Estimating occupancy and abundance of stream amphibians using environmental DNA from filtered water samples. *Can. J. Fish. Aquat. Sci.* **2013**, *70*, 1123–1130. [CrossRef]
38. Mukherjee, I.; Hodoki, Y.; Nakano, S.I. Seasonal dynamics of heterotrophic and plastidic protists in the water column of Lake Biwa, Japan. *Aquat. Microb. Ecol.* **2017**, *80*, 123–137. [CrossRef]
39. Medlin, L.; Elwood, H.J.; Stickel, S.; Sogin, M.L. The characterization of enzymatically amplified eukaryotic 16S-like rRNA-coding regions. *Gene* **1988**, *71*, 491–499. [CrossRef]
40. Larkin, M.A.; Blackshields, G.; Brown, N.P.; Chenna, R.; McGettigan, P.A.; McWilliam, H.; Valentin, F.; Wallace, I.M.; Wilm, A.; Lopez, R.; et al. Clustal W and Clustal X version 2.0. *Bioinformatics* **2007**, *23*, 2947–2948. [CrossRef]
41. Altschul, S.F.; Gish, W.; Miller, W.; Myers, E.W.; Lipman, D.J. Basic local alignment search tool. *J. Mol. Biol.* **1990**, *215*, 403–410. [CrossRef]
42. Moro, C.V.; Crouzet, O.; Rasconi, S.; Thouvenot, A.; Coffe, G.; Batisson, I.; Bohatier, J. New design strategy for development of specific primer sets for PCR-based detection of Chlorophyceae and Bacillariophyceae in environmental samples. *Appl. Environ. Microbiol.* **2009**, *75*, 5729–5733. [CrossRef]
43. Zimmermann, J.; Jahn, R.; Gemeinholzer, B. Barcoding diatoms: evaluation of the V4 subregion on the 18S rRNA gene including new primers and protocols. *Org. Divers. Evol.* **2011**, *11*, 173–192. [CrossRef]
44. Taton, A.; Grubisic, S.; Brambilla, E.; De Wit, R.; Wilmotte, A. Cyanobacterial diversity in natural and artificial microbial mats of Lake Fryxell (McMurdo Dry Valleys, Antarctica): a morphological and molecular approach. *Appl. Environ. Microbiol.* **2003**, *69*, 5157–5169. [CrossRef]
45. Weisburg, W.G.; Barns, S.M.; Pelletier, D.A.; Lane, D.J. 16S ribosomal DNA amplification for phylogenetic study. *J. Bacteriol.* **1991**, *173*, 697–703. [CrossRef] [PubMed]
46. Kim, Y.W.; Min, B.R.; Choi, Y.K. The genetic diversity of bacterial communities in the groundwater. *Korean J. Environ. Biol.* **2000**, *18*, 53–61.
47. Lara, E.; Berney, C.; Harms, H.; Chatzinotas, A. Cultivation-independent analysis reveals a shift in ciliate 18S rRNA gene diversity in a polycyclic aromatic hydrocarbon-polluted soil. *FEMS Microbiol. Ecol.* **2007**, *62*, 365–373. [CrossRef] [PubMed]

48. Chang, K.H.; Doi, H.; Nishibe, Y.; Nakano, S.I. Feeding habits of omnivorous Asplanchna: comparison of diet composition among *Asplanchna herricki, A. priodonta* and *A. girodi* in pond ecosystems. *J. Limnol.* **2010**, *69*, 209–216. [CrossRef]
49. Kleinow, W. Biochemical studies on Brachionus plicatilis: hydrolytic enzymes, integument proteins and composition of trophi. In *Rotifer Symposium VI.*; Gilbert, J.J., Lubzens, E., Miracle, M.R., Eds.; Springer: Dordrecht, The Nederlands, 1993; pp. 231–246.
50. Kim, K.H.; Park, D.E.; Oh, S. Effects of heat treatment on the nutritional quality of milk: II. destruction of microorganisms in milk by heat treatment. *J. Milk Sci. Biotechnol.* **2017**, *35*, 55–72. [CrossRef]
51. Choi, H.J.; Kim, B.H.; Kim, J.D.; Han, M.S. *Streptomyces neyagawaensis* as a control for the hazardous biomass of *Microcystis aeruginosa* (Cyanobacteria) in eutrophic freshwaters. *Biol. Control* **2005**, *33*, 335–343. [CrossRef]
52. Shunyu, S.; Yongding, L.; Yinwu, S.; Genbao, L.; Dunhai, L. Lysis of *Aphanizomenon flos-aquae* (Cyanobacterium) by a bacterium *Bacillus cereus*. *Biol. Control* **2006**, *39*, 345–351. [CrossRef]
53. Ooms-Wilms, A.L. Are bacteria an important food source for rotifers in eutrophic lakes? *J. Plankton Res.* **1997**, *19*, 1125–1141. [CrossRef]
54. Adl, S.M.; Bass, D.; Lane, C.E.; Lukeš, J.; Schoch, C.L.; Smirnov, A.; Cárdenas, P. Revisions to the classification, nomenclature, and diversity of eukaryotes. *J. Eukaryot. Microbiol.* **2019**, *66*, 4–119.
55. Simek, K.; Jezbera, J.; Horňak, K.; Vrba, J.; Seda, J. Role of diatom-attached choanoflagellates of the genus Salpingoeca as pelagic bacterivores. *Aquat. Microb. Ecol.* **2004**, *36*, 257–269. [CrossRef]
56. Frenken, T.; Wierenga, J.; van Donk, E.; Declerck, S.A.; de Senerpont Domis, L.N.; Rohrlack, T.; Van de Waal, D.B. Fungal parasites of a toxic inedible cyanobacterium provide food to zooplankton. *Limnol. Oceanogr.* **2018**, *63*, 2384–2393. [CrossRef]

Article

Discrimination of Spatial Distribution of Aquatic Organisms in a Coastal Ecosystem Using eDNA

Hyunbin Jo [1], Dong-Kyun Kim [1], Kiyun Park [1] and Ihn-Sil Kwak [1,2,*]

[1] Fisheries Science Institute, Chonnam National University, Yeosu 59626, Korea
[2] Faculty of Marine Technology, Chonnam National University, Yeosu 59626, Korea
* Correspondence: inkwak@hotmail.com; Tel.: +82-61-659-7148

Received: 4 July 2019; Accepted: 15 August 2019; Published: 21 August 2019

Featured Application: Complement monitoring tool.

Abstract: The nonlinearity and complexity of coastal ecosystems often cause difficulties when analyzing spatial and temporal patterns of ecological traits. Environmental DNA (eDNA) monitoring has provided an alternative to overcoming the aforementioned issues associated with classical monitoring. We determined aquatic community taxonomic composition using eDNA based on a meta-barcoding approach that characterizes the general ecological features in the Gwangyang Bay coastal ecosystem. We selected the V9 region of the 18S rDNA gene (18S V9), primarily because of its broad range among eukaryotes. Our results produced more detailed spatial patterns in the study area previously categorized (inner bay, main channel of the bay and outer bay) by Kim et al. (2019). Specifically, the outer bay zone was clearly identified by CCA using genus-level identification of aquatic organisms based on meta-barcoding data. We also found significant relationships between environmental factors. Therefore, eDNA monitoring based on meta-barcoding approach holds great potential as a complemental monitoring tool to identify spatial taxonomic distribution patterns in coastal areas.

Keywords: coastal ecosystem; eDNA; spatial patterns; complemental monitoring tool

---

1. Introduction

Biological monitoring contributes to the understanding of complex ecosystem structures and functions in targeted systems [1]. Accordingly, it is crucial in detecting and assessing environmental changes in order to ensure proper management and conservation of complex ecosystems [2]. Coastal environments are among the most complex ecosystems due to tidal activity, and typically retain high economic and environmental values in light of aquatic resources and biodiversity [3]. Coastal ecosystems are often severely affected by anthropogenic activities such as industrial fishing [4,5], marine transport and leisure activities [6], aquaculture and the aquarium trade [7,8], living seafood and lure fisheries [9,10], and non-indigenous species (NIS) which induce greater pressures on endemic ecosystems and often drive native species to extinction in the resulting habitats [11–13]. The nonlinearity and complexity of coastal ecosystems due to the aforementioned activities often causes difficulties when analyzing spatial and temporal patterns of ecological traits.

Environmental DNA (eDNA) as described by Ogram et al. [14] who extracted microbial DNA from the sediment, has increasingly been used in recent years for biological monitoring purposes. Recently, a large number of papers have reported the use of eDNA monitoring in analyses of soil, water and even air [15]. Andersen et al. [16] examined the possibility of monitoring large mammals using eDNA in soil samples, and eDNA from water monitoring of fish [17–20] and amphibians [21,22] has been successful. Furthermore, Hawkins et al. [23] demonstrated that a complete taxonomic list of functional feeding group (FFG) criteria, based on high resolution of identification (genus or species level) based on DNA techniques, can determine the effects of watershed alterations on stream invertebrate assemblages in bulk eDNA samples. However, family level identification based on visual inspection did not reveal any differences of FFG composition between sites. Next-generation sequencing (NGS) technologies for eDNA monitoring have provided an alternative to overcome issues such as identification problems associated with classical monitoring in a species rich coastal environment [3,20].

The values of coastal ecosystems, such as primary production (i.e., sea grass and algae) and commercial fish yields, are intertwined with multiple environmental factors, including nutrient concentrations (carbon, nitrogen, and phosphorus), phytoplankton growth, zooplankton grazing effects, and benthic communities. The Gwangyang Bay coastal ecosystem is the most economically productive in Korean peninsula. Specifically, in the midst of three major cities (Gwangyang, Yeosu and Suncheon) in Jeonnam Province, with Gwangyang Bay, it yields 71% (1,297,815 tons) of the annual aquacultural resource output as of 2016 [24]. However, there is a large industrial area near the Bay, and the area is primarily involved in industrial activities such as oil refineries and steel production plants. Kim et al. [25] characterized the dissimilarity of water quality and sediment contamination, and identified the importance of nutrients supplied by rivers. Such findings are, however, still limited to representing general ecological features of the Gwangyang Bay coastal ecosystem.

The main objective of this study was to determine aquatic community taxonomic composition using eDNA based on an NGS approach for characterizing general ecological features in the Gwangyang Bay coastal ecosystem. We analyzed the community spatial distribution with regard to environmental parameters, and the habitat types (marine, freshwater and estuarine), feeding habits (filter feeder, carnivore, producer and symbiotic) and indigenous species rate (ISR) among the three different zones referred to by Kim et al. [25]. Moreover, we discuss the effectiveness and sensitivity of our NGS approach on the Gwangyang Bay coastal ecosystem.

2. Materials and Methods

2.1. Study Area

Gwangyang Bay is part of the Korean National Archipelagos located off the south coast of the Korean peninsula (Figure 1). The bay receives an annual mean discharge of 2298×10^6 m^3 yr^{-1} from the Seomjin River [26]. A significant amount of nutrients drains into the system from the watershed (~5000 km^2). The water depth varies from 10 m at the Seomjin River estuary to 50 m at the outer Gwangyang Bay. The tidal cycle appears to be semi-diurnal. Compared to other Korean river estuaries which have barriers, the Seomjin River estuary remains open, and thus the water mass is exchanged between the river and ocean more actively. The natural condition of Gwangyang Bay is apt to increase primary productivity as well as biological diversity. In this respect, Gwangyang Bay (~450 km^2 from the estuary to the outer bay) is the most economically productive coastal ecosystem in the Korean peninsula [25].

2.2. Sampling, Data Collection and Primer Selection

A survey permitted by Ha-dong local government (permission number: 2010-0165) was conducted in June 2018. We sampled the surface water (approximately the top 50 cm) in this study. In total, nineteen sampling sites were selected, which covered an extensive area from the Seomjin River estuary to the outer Gwangyang Bay (Figure 1). According to Kim et al. [25], at Gwangyang Bay, higher water temperatures corresponded to lower salinity, and vice versa. Phosphorus and nitrogen concentrations were spatially similar across the Bay. Based on the divisions indicated in Kim et al. [25], Zone I covered the inner Bay (sites 3–7, and 9), Zone II represented the main channel of the Bay (sites 8, 10, and 11), and Zone III mostly belonged to the outer Bay (sites 12–21) (Figure 1). Water samples for meta-barcoding analysis (more than 1 L per sample) were obtained at the same time and moved with dry ice to laboratory to filter (0.45 μm pore-size membrane; Advantec MFS membrane filter, Dublin, USA) and stored at −80 °C before NGS analysis. Negative controls were included for every study site to prevent cross contamination. Water temperature and salinity were measured on-site using portable equipment (Model: YSI Professional Plus, OH, USA), while the nutrient and chlorophyll-*a* concentrations were analyzed in the lab. Phosphorus and nitrogen concentrations were measured using an UV spectrophotometer based on standard analytical methods proposed by the Korean Ministry of Oceans and Fisheries. Chlorophyll-*a* measurements were also based on UV spectrophotometry. In contrast to nutrient measurements, chlorophyll-*a* samples were filtered through a 0.45 μm pore-size membrane (Model: Advantec MFS membrane filter). The filter membrane was then homogenized after acetone extraction prior to spectrophotometry. Organic and inorganic carbon concentrations were measured using a carbon analyzer (Model: vario TOC cub, Langenselbold, Germany) using 850 °C combustion catalytic oxidation methods.

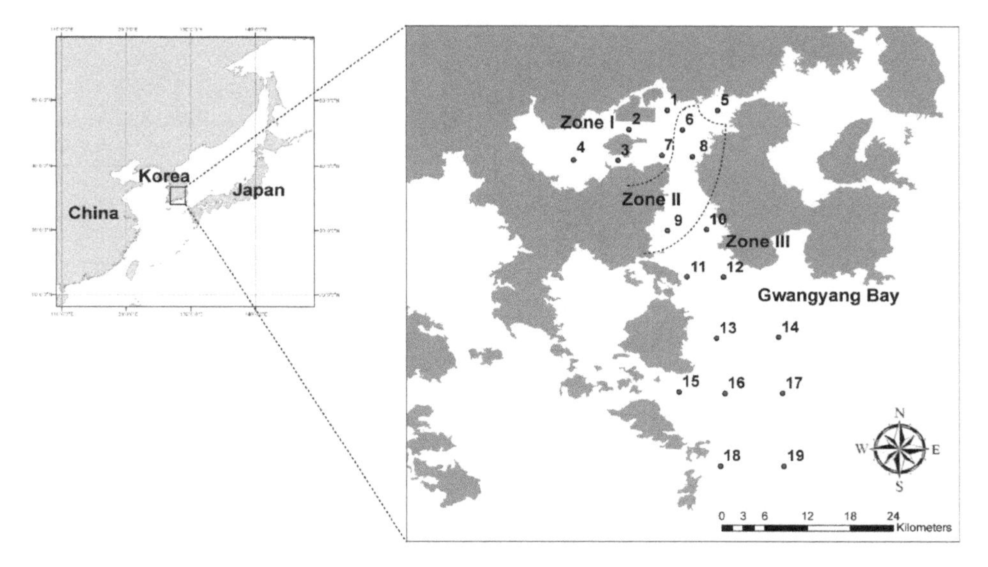

Figure 1. Map of the study sites with division of zones referred to by Kim et al. (2019) [25].

We selected the V9 region of the 18S rDNA gene (18S V9), primarily because of its broad range among eukaryotes [27,28]. NGS approaches using the 18S V9 region have recently allowed the characterization of marine planktonic biodiversity in the oceans [29] and prompted biomarker establishment initiatives [30].

2.3. DNA Extraction and Metagenomic Sequencing

Genomic DNA was extracted using PowerSoil® DNA Isolation Kit (Cat. No. 12888, Qiagen, Düsseldorf, Germany) according to the manufacturer's protocol. DNA extracted for sequencing was prepared according to Illumina 18S Metagenomic Sequencing Library protocols (San Diego, CA, USA). DNA quantity, quality, and integrity were measured by PicoGreen (Thermo Fisher Scientific, Waltham, MA, USA) and VICTOR Nivo Multimode Microplate Reader (PerkinElmer, Akron, OH, USA). The 18S rRNA gene was amplified using 18S V9 primers. The primer sequences are as follows: 18S V9 primer including adaptor sequence (Forward Primer: 5′ TCGTCGGCAGCGTCAGATGTGTATAA GAGACAGCCCTGCCHTTTGTACACAC 3′/Reverse Primer: 5′ GTCTCGTGGGCTCGGAGATG TGTATAAGAGACAGCCTTCYGCAGGTTCACCTAC 3′). To amplify the target region attached with adapters, as a first PCR process, the extracted DNA was amplified by 18S V9 primers with one cycle of 3 min at 95 °C, 25 cycles of 30 s at 95 °C, 30 s at 55 °C, 30 s at 72 °C, and a final step of 5 min at 72 °C for amplicon PCR product. As a second process, to produce indexing PCR, the first PCR product was subsequently amplified with one cycle of 3 min at 95 °C, eight cycles of 30 s at 95 °C, 30 s at 55 °C, 30 s at 72 °C, and a final step of 5 min at 72 °C. The final products were normalized and pooled using PicoGreen (Thermo Fisher Scientific, Waltham, MA, USA), and the size of libraries were verified using the LabChip GX HT DNA High Sensitivity Kit (PerkinElmer, Akron, OH, USA).

The library was sequenced using the MiSeq™ NGS platform (Illumina, San Diego, CA, USA) provided as a commercial service (Macrogen Inc., Seoul, Korea). Raw reads were trimmed with CD-HIT-OTU and chimeras were identified and removed using rDnaTools. For paired-end merging, FLASH (Fast Length Adjustment of SHort reads) version 1.2.11 was used. Merged reads were processed using Qiime version 1.9 [31] and were clustered into operational taxonomic units (OTUs) with UCLUST [32], using a greedy algorithm with OTUs at a 97% OUT cutoff value. Taxonomic classifications were assigned to the obtained representative sequences using BLASTn [33] and UCLUST [32].

2.4. Data Analysis and Statistics

OTUs assigned by meta-barcoding were classified into habitat types (marine, freshwater and estuarine), trophic level (first consumer [filter feeder], second consumer [carnivore], producer and symbiotic) and ISR (indigenous, non-indigenous). Canonical correspondence analysis (CCA) of OTUs exhibiting >1% relative abundance was performed using the PAST 3.0 program [34] to evaluate relationships between abundance of OTU sequences (based on genus level identification) and environmental factors (temperature (Temp.), salinity, total phosphorus (TP), total nitrogen (TN), chlorophyll-*a* (Chl-*a*), total organic carbon (TOC), total inorganic carbon (TIC), total carbon (TO), elemental carbon (EC)). All CCA results were constructed using relative abundance data, with natural logarithms transformation (ln 1 + X) used for sample normalization. Linear relationships (Pearson correlations) were calculated between the above environmental factors and classified OTU sequences, based on habitat using XLSTAT version 2018.6.54467 (64 bit) as a plug-in for the Microsoft Excel program [35].

3. Results

3.1. Meta-Barcoding

In total, 3,066,013 paired-end reads from the 19 samples were generated on the Illumina MiSeq™ platform, of which 98.5% passed Q30 (Phred quality score > 30) for improving accuracy of sequences in this study (Supplementary Materials). Each sample yielded paired-end reads ranging from 21,101–299,305 reads (mean: 161,369 reads), which was similar to the amount of reads in the previous study [36], and all samples exhibited saturation of the number of OTUs by rarefaction curve analysis. Gamma-diversity was 352 OTUs produced with a cutoff of 97% similarity. The resulting 352 OTUs were classified into 19 genus-level taxonomic groups (those representing <0.04% abundance were not plotted). Uncultured and non-assigned reads were discarded.

3.2. Spatial Distributions of Aquatic Organisms Based on Meta-Barcoding

We carried out a survey in June 2018. Relationships among the environmental variables and aquatic organisms based on meta-barcoding were explored by CCA (Figure 2). The first axis explained 33.3% of the variance and distinguished Zones III-2 and III-3 with higher salinity, EC, TIC, and lower Chl-a and TN from the other zones. The second axis (19.6% of variance explained) distinguished the sites from Zone III-1 and the other zones, by having higher TOC and TC, lower temperature, and TP. Nutrient-related factors (Chl-*a*, TN, and TP) were correlated with *Acropora* sp. (small polyp stony coral) and *Megabalanus* sp., (barnacle) and salinity was correlated with *Skeletonema* sp. (diatoms) and *Centropages* sp. (copepods).

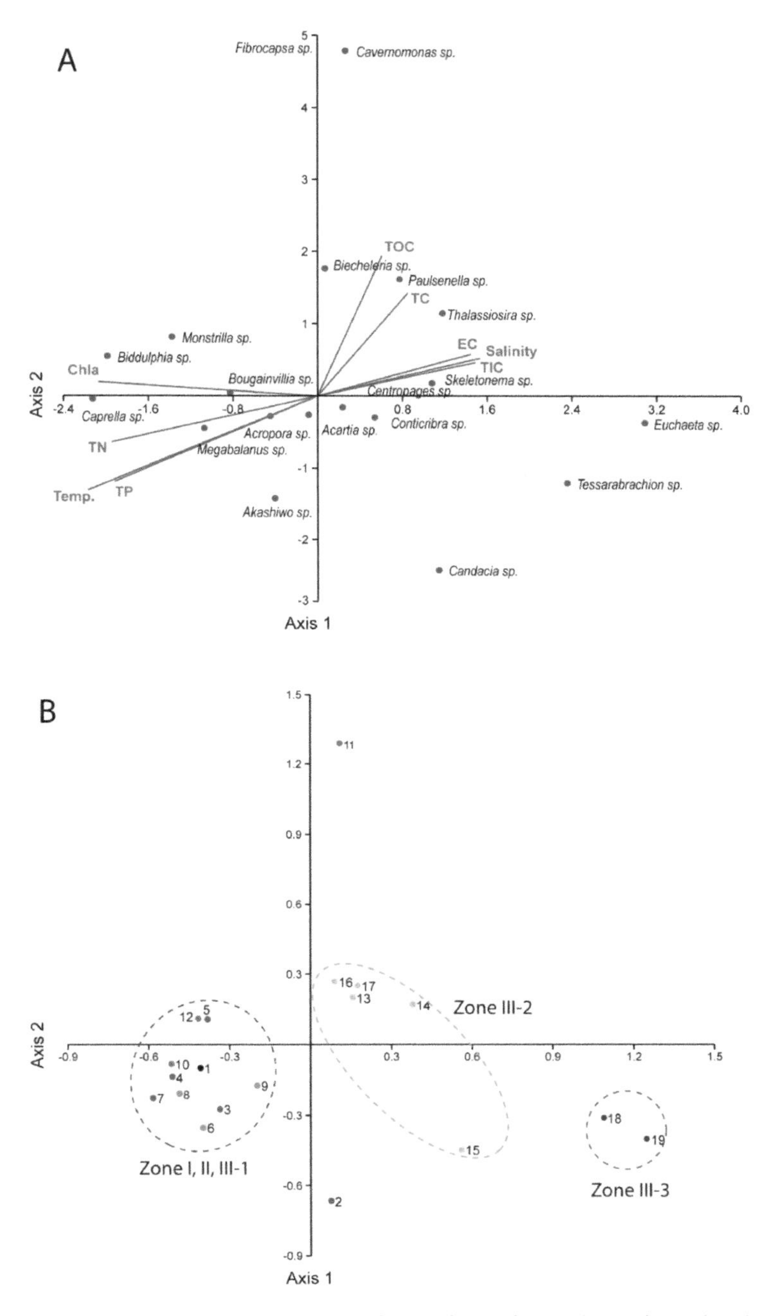

Figure 2. Canonical correspondence analysis (CCA) used to evaluate (**A**) the relationships between abundance of operational taxonomic units (OTUs) sequences (based on genus level identification) and environmental factors (temperature (Temp.), Salinity, total phosphorus (TP), total nitrogen (TN), chlorophyll-a (Chl-a), total organic carbon (TOC), total inorganic carbon (TIC), total carbon (TO), elemental carbon (EC)). (**B**) Sites grouping based on CCA analysis.

Abundance of assigned OTU sequences showed different patterns among the study sites (Figure 3). The dominant OTU was assigned to *Acropora* sp. and the sub-dominant OTU was *Acartia* sp. (copepods). Most common OTUs within the study sites were comprised of phytoplankton, followed by zooplankton and amphipods (Figure 3B). Interestingly, the abundance of OTUs showed different compositions among the three different zones (I, II, III). Zone I showed similar patterns to Zones II and III-1, whereas Zones III-2 and 3 showed different compositions of aquatic organisms which distinguished them from other zones. In particular, Zone III-3 exhibited an entirely different pattern, comprised of marine organisms such as *Euchaeta* sp. (copepods) and *Tessarabrachion* sp. (krill) compared with other zone divisions (Figure 3B).

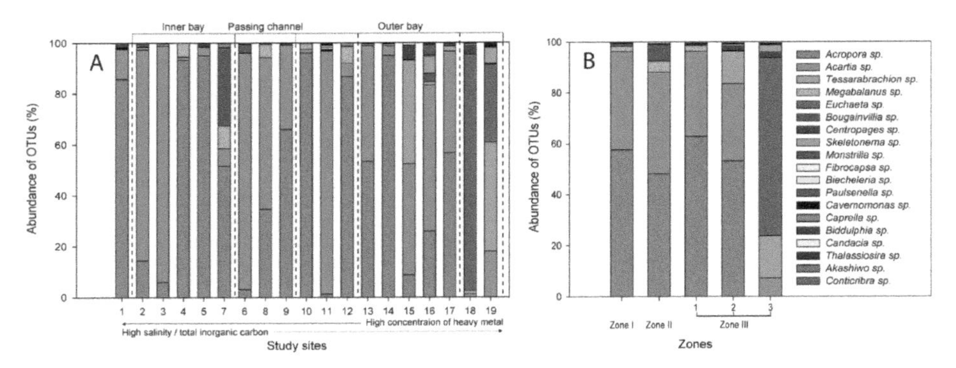

Figure 3. Abundance of OTU sequences along the study sites (**A**) with site description of physical features and zone classification (**B**) based on genus identification level.

3.3. Functional Features and Non-Indigenous Species

When divided into categories (habitat types, trophic level, and ISR), averages of abundant sequences showed different patterns among the zones (I, II, III-1-2-3). The most dominant habitat type of OTU was the marine type for the three different zones (Figure 4A), while the types of feeding habit and indigenous rate showed complex response patterns. The dominant trophic level was first consumer and second consumer, followed by producer and symbiotic trophic levels (Figure 4B). The indigenous rate of Zone III-3 was significantly higher than other zones, but Zones II and III-1 showed opposite patterns (Figure 4C), while Zones I and III-2 presented no significant difference in indigenous rates. The three divisions of Zone III from CCA results demonstrated similar patterns within the three zone types (Zones I, II-and III).

We found significant relationships between environmental factors (Table 1). Specifically, temperature had positive relationships with nutrient-related factors such as TP and PN, and showed a negative relationship with TOC. Salinity showed significant negative relationships with TN and TP, but also displayed a positive relationship with carbon-related factors such as EC and TC. However, there were no significant relationships between environmental factors and divisions of categories except for habitat types (marine, estuarine, freshwater). The marine habitat of OTUs revealed a significant positive relationship with salinity (r = 0.879) and EC (r = 0.878), respectively, whereas freshwater and estuarine types of OTUs showed negative relationships with salinity (r = −0.888 and −0.856) and EC (r = −0.884 and −0.856) respectively.

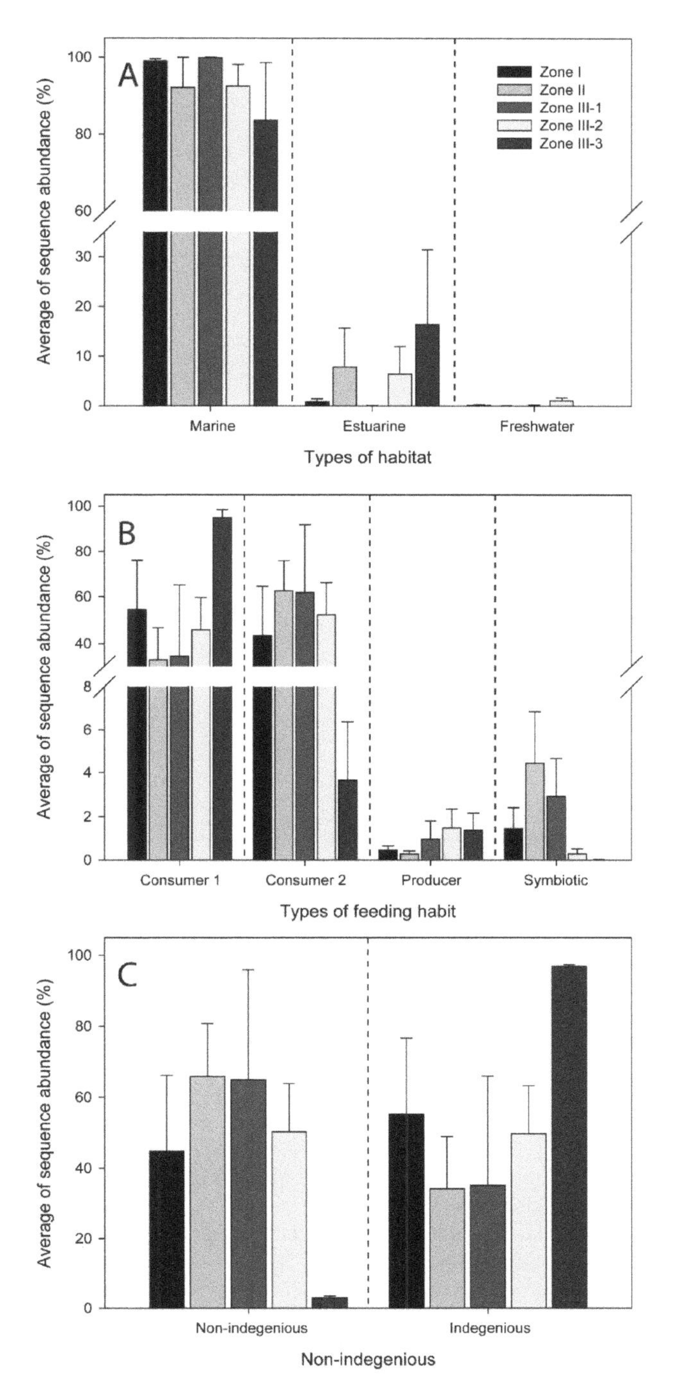

Figure 4. Comparison of the habitat types ((**A**), marine-freshwater-estuarine), trophic level ((**B**), consumer 1 [filter feeder], consumer 2 [carnivore], producer and symbiotic) and natives ((**C**), non-indigenous and indigenous) among five zoning types based on genus identification level.

Table 1. Linear relationship (Pearson correlation) between environmental factors (temperature (Temp.), Salinity, total phosphorus (TP), total nitrogen (TN), chlorophyll-a (Chl-a), total organic carbon (TOC), total inorganic carbon (TIC), total carbon (TO), elemental carbon (EC)) and classified OTU sequences based on habitat types, feeding habits and natives (Bold: significant relationship).

Variables	Temp	Salinity	EC	TP	TN	TOC	TIC	TC	Chla
Temp	**1**	−0.192	−0.216	**0.505**	**0.464**	**−0.599**	−0.275	−0.220	0.422
Salinity	−0.192	**1**	**0.996**	**−0.586**	**−0.532**	0.301	0.435	**0.486**	0.006
EC	−0.216	**0.996**	1	**−0.615**	**−0.564**	0.325	0.440	**0.478**	−0.034
TP	**0.505**	**−0.586**	**−0.615**	1	**0.904**	**−0.505**	**−0.634**	**−0.715**	0.192
TN	**0.464**	**−0.532**	**−0.564**	**0.904**	1	−0.424	**−0.680**	**−0.645**	0.200
TOC	**−0.599**	0.301	0.325	**−0.505**	−0.424	**1**	0.127	0.355	−0.248
TIC	−0.275	0.435	0.440	**−0.634**	**−0.680**	0.127	**1**	**0.877**	0.195
TC	−0.220	**0.486**	**0.478**	**−0.715**	**−0.645**	0.355	**0.877**	**1**	0.263
Chl−*a*	0.422	0.006	−0.034	0.192	0.200	−0.248	0.195	0.263	**1**
Marine	−0.197	**0.879**	**0.878**	−0.343	−0.195	0.314	0.051	0.151	−0.082
Fresh	0.123	**−0.888**	**−0.884**	0.332	0.199	−0.286	−0.061	−0.175	0.024
Estuarine	0.269	**−0.856**	**−0.856**	0.349	0.188	−0.337	−0.041	−0.123	0.139
First consumer	0.195	−0.283	−0.271	0.164	0.120	0.152	−0.329	−0.269	0.159
Second consumer	0.011	0.387	0.373	−0.329	−0.070	0.159	0.239	0.346	0.026
Producer	−0.054	−0.319	−0.307	0.288	0.040	−0.194	−0.162	−0.281	−0.064
Symbiotic	0.032	−0.335	−0.326	0.266	0.182	0.158	−0.298	−0.301	0.220
Non-indigenous	−0.002	−0.348	−0.331	0.284	0.021	−0.137	−0.195	−0.295	−0.008
Indigenous	0.002	0.348	0.331	−0.284	−0.021	0.137	0.195	0.295	0.008

4. Discussion

4.1. *Effectiveness of eDNA Monitoring*

Our results showed that eDNA monitoring based on NGS holds great potential as a complementary monitoring tool to identify spatial taxonomic distribution patterns in coastal areas. We characterized detailed zonation patterns of the outer bay of Gwangyang Bay previously categorized by Kim et al. [25]. Specifically, Zones III-2 and III-3 were clearly resolved by CCA using genus-level identification of aquatic organisms based on meta-barcoding data (Figure 2). When divided into categories (habitat types, trophic level, and ISR), averages of abundant sequences showed different patterns among the zones (I, II, III-1-2-3). The NIS detection of Zone III-3 was significantly higher than the other zones. Therefore, eDNA based on meta-barcoding is a promising ecological tool for monitoring, such as biodiversity assessment or NIS management [24,37–40].

This approach by meta-barcoding in complex coastal ecosystems confers three distinct advantages over other methods. First, the range of aquatic organisms in coastal ecosystems open to study is widened. Second, identification of aquatic organism to the species or genus level (Appendix A), which can be converted into ecological values such as FFC and NIS, is possible. Finally, sensitivity of the meta-barcoding approach using small volumes of water (1 L) is a promising alternative to traditional methods (i.e., dredges and surber samplers) for biodiversity detection in coastal areas. Using these advantages, eDNA monitoring can provide a useful tool for use in environmental management.

However, some taxa were unable to be identified to the species or genus level due to the incompleteness of reference databases (i.e., NCBI GenBank). More accurate target regions such as the cytochrome oxidase I region, which is called the standard region of barcoding, are needed to identify target species at the species level and are required for identification of aquatic organisms [41]. Another challenge associated with eDNA monitoring is the risk of false-positive and false-negative detections [42]. The reliability of the eDNA sampling method should be demonstrated using in silico, in vitro and in situ validation tests in the coastal area. Even though this approach has limitations, taxonomic expertise is not required and it can supplement observational records and field surveys to obtain marine ecosystem samples and information, as NGS reveals biodiversity and the number of NIS in one specific area along the their temporal and spatial distribution.

4.2. Ecological Values of eDNA Monitoring

Our results showed that *Acropora* sp. was dominant in terms of abundance of OTU sequences across all study sites, even though it is not a planktonic species. The Acropora colonies post-recruitment by larval recruitment demonstrates higher efficiency for survival than coral-colony growth [43]. It was concluded that larval recruitment largely determines species composition, and that reduced larval recruitment is responsible for the sparse distribution of fragmenting species [44]. Therefore, the larval stage could be easily detected among the study sites by our eDNA monitoring approach. We also found that nutrient-related factors (Chl-a, TN, and TP) were significantly correlated with *Acropora* sp. and *Megabalanus* sp. (Figure 2). It is therefore possible that the strong positive correlation between the two species could be the result of coral-inhabiting barnacles [45].

The comprehensive understanding of feeding characteristics of aquatic organisms has not been well elucidated in comparison to their importance [46–49]. Our results showed that composition of aquatic organisms have different patterns of feeding habits among the three different zone divisions designated by Kim et al. [25] (Figure 4), and we also found similar patterns using CCA based on genus-level identification resulting from our division of Zone III (Figure 2). However, previous research findings were attributed to the absence of adequate information for analysis obtained from the field sites due to difficulties in culture, handling, and identification of eDNA samples derived from bulk sediment and filtered water. In this sense, our results overcame the aforementioned limitations by using a broader detection range for aquatic organisms.

Although, Zhan et al. [50] has described the increased sensitivity of meta-barcoding for NIS, its application for monitoring biological invasion has only recently been demonstrated [51]. In the present work, we identified the utility of meta-barcoding for detection of NIS and their spatial distribution patterns (Figure 4). The reason behind this is the capacity to detect the presence of individuals at early life stages, such as eggs or nauplius larvae, whose identification is difficult with traditional methods [52]. The sensitivity of meta-barcoding, combined with the relatively low time and cost associated with this technique [53], makes it a promising alternative approach for the rapid and accurate detection of biodiversity shifts in aquatic organisms, allowing its potential implementation in environmental policies.

Supplementary Materials: DNA sequences: DRYAD entry https://datadryad.org/review?doi=doi:10.5061/dryad.41b1dp3.

Author Contributions: Conceptualization, H.J., D.-K.K. and I.-S.K.; methodology, H.J. and K.P.; formal analysis, H.J. and D.-K.K.; investigation, H.J. and D.-K.K.; writing—original draft preparation, H.J. and I.-S.K.; writing—review and editing, D.-K.K. and I.-S.K.; project administration, I.-S.K.; funding acquisition, I.-S.K.

Funding: This research was funded by the National Research Foundation of Korea, grant number NRF-2018R1A6A1A03024314.

Conflicts of Interest: The authors declare no conflict of interest.

Appendix A Species Level Identification List Used in this Paper.

Kingdom	Species
Eukaryota	*Acropora granulosa*
Eukaryota	Acartia omorii
Eukaryota	*Tessarabrachion oculatum*
Eukaryota	*Megabalanus stultus*
Eukaryota	*Euchaeta indica*
Eukaryota	*Bougainvillia muscus*
Eukaryota	*Centropages typicus*
Eukaryota	*Skeletonema costatum*
Eukaryota	*Monstrilla sp.*
Eukaryota	*Fibrocapsa japonica*
Eukaryota	*Biecheleria brevisulcata*
Eukaryota	*Paulsenella vonstoschii*

Kingdom	Species
Eukaryota	*Cavernomonas mira*
Eukaryota	*Caprella californica*
Eukaryota	*Biddulphia sp.*
Eukaryota	*Candacia bispinosa*
Eukaryota	*Thalassiosira mala*
Eukaryota	*Akashiwo sanguinea*
Eukaryota	*Conticribra weissflogiopsis*

References

1. Jo, H.; Ventura, M.; Vidal, N.; Gim, J.S.; Buchaca, T.; Barmuta, L.A.; Erik, J.; Joo, G.J. Discovering hidden biodiversity: The use of complementary monitoring of fish diet based on DNA barcoding in freshwater ecosystems. *Ecol. Evol.* **2016**, *6*, 219–232. [CrossRef] [PubMed]
2. Robertson, G.P.; Collins, S.L.; Foster, D.R.; Brokaw, N.; Ducklow, H.W.; Gragson, T.L.; Gries, C.; Hamilton, S.K.; McGuire, A.D.; Moore, J.C.; et al. Long-term ecological research in a human-dominated world. *BioScience* **2006**, *62*, 342–353. [CrossRef]
3. Baird, D.J.; Hajibabaei, M. Biomonitoring 2.0: A new paradigm in ecosystem assessment made possible by next-generation DNA sequencing. *Mol. Ecol.* **2012**, *21*, 2039–2044. [CrossRef] [PubMed]
4. Myers, R.A.; Worm, B. Rapid worldwide depletion of predatory fish communities. *Nature* **2003**, *423*, 280. [CrossRef] [PubMed]
5. Pusceddu, A.; Bianchelli, S.; Martín, J.; Puig, P.; Palanques, A.; Masqué, P.; Danovaro, R. Chronic and intensive bottom trawling impairs deep-sea biodiversity and ecosystem functioning. *Proc. Natl. Acad. Sci. USA* **2014**, *111*, 8861–8866. [CrossRef] [PubMed]
6. Ruiz, G.M.; Fofonoff, P.W.; Carlton, J.T.; Wonham, M.J.; Hines, A.H. Invasion of coastal marine communities in North America: Apparent patterns, processes, and biases. *Annu. Rev. Ecol. Syst.* **2000**, *31*, 481–531. [CrossRef]
7. Naylor, R.; Williams, S.L.; Strong, D.R. Aquaculture—A gateway for exotic species. *Science* **2001**, *294*, 1655–1656. [CrossRef]
8. Padilla, D.K.; Williams, S.L. Beyond ballast water: Aquarium and ornamental trades as sources of invasive species in aquatic ecosystems. *Front. Ecol. Environ.* **2004**, *2*, 131–138. [CrossRef]
9. Chapman, J.W.; Miller, T.W.; Coan, E.V. Live seafood species as recipes for invasion. *Conserv. Biol.* **2003**, *17*, 1386–1395. [CrossRef]
10. Weigel, S.; Bester, K.; Hühnerfuss, H. Identification and quantification of pesticides, industrial chemicals, and organobromine compounds of medium to high polarity in the North Sea. *Mar. Pollut. Bull.* **2005**, *50*, 252–263. [CrossRef]
11. Ardura, A.; Planes, S. Rapid assessment of non-indigenous species in the era of the eDNA barcoding: A Mediterranean case study. *Estuar. Coast. Shelf Sci.* **2017**, *188*, 81–87. [CrossRef]
12. Carlton, J.T.; Geller, J.B. Ecological roulette: The global transport of nonindigenous marine organisms. *Science* **1993**, *261*, 78–82. [CrossRef]
13. Williams, S.L.; Grosholz, E.D. The invasive species challenge in estuarine and coastal environments: Marrying management and science. *Estuaries Coasts* **2008**, *31*, 3–20. [CrossRef]
14. Ogram, A.; Sayler, G.S.; Barkay, T. The extraction and purification of microbial DNA from sediments. *J. Microbiol. Meth.* **1987**, *7*, 57–66. [CrossRef]
15. Taberlet, P.; Coissac, E.; Hajibabaei, M.; Rieseberg, L.H. Environmental DNA. *Mol. Ecol.* **2012**, *21*, 1789–1793. [CrossRef]
16. Andersen, K.; Bird, K.L.; Rasmussen, M.; Haile, J.; Breuning-Medsen, H.; Kjaer, K.H.; Orlando, L.; Geilbert, M.T.P.; Willerslev, E. Meta-barcoding of 'dirt' DNA from soil reflects vertebrate biodiversity. *Mol. Ecol.* **2011**, *21*, 1966–1979. [CrossRef]
17. Minamoto, T.; Yamanaka, H.; Takahara, T.; Honjo, M.N.; Kawabata, Z. Surveillance of fish species composition using environmental DNA. *Limnology* **2012**, *13*, 193–197. [CrossRef]

18. Thomsen, P.; Kielgast, J.; Iversen, L.L.; Wiuf, C.; Rasmussen, M.; Gilbert, M.T.P.; Orlando, L.; Willerslev, E. Monitoring endangered freshwater biodiversity using environmental DNA. *Mol. Ecol.* **2012**, *21*, 2565–2573. [CrossRef]
19. Thomsen, P.; Kielgast, J.; Iversen, L.L.; Møller, P.R.; Rasmussen, M.; Willerslev, E. Detection of a diverse marine fish fauna using environmental DNA from seawater samples. *PLoS ONE* **2012**, *7*, e41732. [CrossRef]
20. Yamamoto, S.; Masuda, R.; Sato, Y.; Sado, T.; Araki, H.; Kondoh, M.; Toshifumi, M.; Miya, M. Environmental DNA metabarcoding reveals local fish communities in a species-rich coastal sea. *Sci. Rep.* **2017**, *7*, 40368. [CrossRef]
21. Ficetola, G.F.; Miaud, C.; Pompanon, F.; Taberlet, P. Species detection using environmental DNA from water samples. *Biol. Lett.* **2008**, *4*, 423–425. [CrossRef]
22. Goldberg, C.S.; Pilliod, D.S.; Arkle, R.S.; Waits, L.P. Molecular detection of vertebrates in stream water: A demonstration using Rocky Mountain tailed frogs and Idaho giant salamanders. *PLoS ONE* **2011**, *6*, e22746. [CrossRef]
23. Hawkins, C.P.; Norris, R.H.; Hogue, J.N.; Feminella, J.W. Development and evaluation of predictive models for measuring the biological integrity of streams. *Ecol. Appl.* **2000**, *10*, 1456–1477. [CrossRef]
24. Korean Statistical Information Service (KOSIS). Available online: http://kosis.kr (accessed on 9 June 2019).
25. Kim, D.K.; Jo, H.; Han, I.; Kwak, I.S. Explicit Characterization of Spatial Heterogeneity Based on Water Quality, Sediment Contamination, and Ichthyofauna in a Riverine-to-Coastal Zone. *Int. J. Environ. Res. Health* **2019**, *16*, 409. [CrossRef]
26. Kang, C.K.; Kim, J.B.; Lee, K.S.; Kim, J.B.; Lee, P.Y.; Hong, J.S. Trophic importance of benthic microalgae to macrozoobenthos in coastal bay systems in Korea: Dual stable C and N isotope analyses. *Mar. Ecol. Prog. Ser.* **2003**, *259*, 79–92. [CrossRef]
27. Abad, D.; Albaina, A.; Aguirre, M.; Laza-Martínez, A.; Uriarte, I.; Iriarte, A.; Arantza, I.; Fernando, V.; Estonba, A. Is metabarcoding suitable for estuarine plankton monitoring? A comparative study with microscopy. *Mar. Biol.* **2016**, *163*, 149. [CrossRef]
28. de Vargas, C.; Audic, S.; Henry, N.; Decelle, J.; Mahé, F.; Logares, R.; Lara, E.; Berney, C. Eukaryotic plankton diversity in the sunlit ocean. *Science* **2015**, *348*, 1261605. [CrossRef]
29. Albaina, A.; Aguirre, M.; Abad, D.; Santos, M.; Estonba, A. 18S rRNA V9 metabarcoding for diet characterization: A critical evaluation with two sympatric zooplanktivorous fish species. *Ecol. Evol.* **2016**, *6*, 1809–1824. [CrossRef]
30. Massana, R.; Gober, A.; Audic, S.; Bass, D.; Bittner, L.; Boutte, C.; Chambouvet, A.; Christen, R. Marine protist diversity in European coastal waters and sediments as revealed by high-throughput sequencing. *Environ. Microbiol.* **2015**, *17*, 4035–4049. [CrossRef]
31. Caporaso, J.G.; Kuczynski, J.; Stombaugh, J.; Bittinger, K.; Bushman, F.D.; Costello, E.K.; Huttley, G.A. QIIME allows analysis of high-throughput community sequencing data. *Nat. Methods* **2010**, *7*, 335. [CrossRef]
32. Edgar, R.C. Search and clustering orders of magnitude faster than BLAST. *Bioinformatics* **2010**, *26*, 2460–2461. [CrossRef]
33. Altschul, S.F.; Gish, W.; Miller, W.; Myers, E.W.; Lipman, D.J. Basic local alignment search tool. *J. Mol. Biol.* **1990**, *215*, 403–410. [CrossRef]
34. Hammer, Ø.; Harper, D.A.; Ryan, P.D. PAST: Paleontological statistics software package for education and data analysis. *Palaeontol. Electron.* **2001**, *4*, 9.
35. Addinsoft. *XLSTAT Statistical and Data Analysis Solution*; Addinsoft: New York, NY, USA, 2019.
36. Thomsen, P.F.; Møller, P.R.; Sigsgaard, E.E.; Knudsen, S.W.; Jørgensen, O.A.; Willerslev, E. Environmental DNA from seawater samples correlate with trawl catches of subarctic, deepwater fishes. *PLoS ONE* **2016**, *11*, e0165252. [CrossRef]
37. Bourlat, S.J.; Borja, A.; Gilbert, J.; Taylor, M.I.; Davies, N.; Weisberg, S.B.; Glöckner, F.O. Genomics in marine monitoring: New opportunities for assessing marine health status. *Mar. Pollut. Bull.* **2013**, *74*, 19–31. [CrossRef]
38. Valentini, A.; Taberlet, P.; Miaud, C.; Civade, R.; Herder, J.; Thomsen, P.F.; Bellemain, E.; Besnard, A.; Coissac, E.; Boyer, F.; et al. Next-generation monitoring of aquatic biodiversity using environmental DNA barcoding. *Mol. Ecol.* **2016**, *25*, 929–942. [CrossRef] [PubMed]
39. O'Donnell, J.L.; Kelly, R.P.; Shelton, A.O.; Samhouri, J.F.; Lowell, N.C.; Williams, G.D. Spatial distribution of environmental DNA in a nearshore marine habitat. *Peer J.* **2017**, *5*, e3044. [CrossRef] [PubMed]

40. Djurhuus, A.; Pitz, K.; Sawaya, N.A.; Rojas-Márquez, J.; Michaud, B.; Montes, E.; Muller-Karger, F.; Breitbart, M. Evaluation of marine zooplankton community structure through environmental DNA metabarcoding. *Limnol. Oceanogr. Methods* **2018**, *16*, 209–221. [CrossRef]
41. Hajibabaei, M.; Smith, M.A.; Janzen, D.H.; Rodriguez, J.J.; Whitfield, J.B.; Hebert, P.D. A minimalist barcode can identify a specimen whose DNA is degraded. *Mol. Ecol. Notes* **2006**, *6*, 959–964. [CrossRef]
42. Darling, J.A.; Mahon, A.R. From molecules to management: Adopting DNA-based methods for monitoring biological invasions in aquatic environments. *Environ. Res.* **2011**, *111*, 978–988. [CrossRef]
43. Victor, S.; Golbuu, Y.; Yukihira, H.; van Woesik, R. Acropora size-frequency distributions reflect spatially variable conditions on coral reefs of Palau. *Bull. Mar. Sci.* **2009**, *85*, 149–157.
44. Wallace, C.C. Reproduction, recruitment and fragmentation in nine sympatric species of the coral genus Acropora. *Mar. Biol.* **1985**, *88*, 217–233. [CrossRef]
45. Lewis, J.B. Recruitment, growth and mortality of a coral-inhabiting barnacle *Megabalanus stultus* (Darwin) upon the hydrocoral *Millepora complanata* Lamarck. *J. Exp. Mar. Biol. Ecol.* **1992**, *162*, 51–64. [CrossRef]
46. Rothhaupt, K. Differences in particle size-dependent feeding efficiencies of closely related rotifer species. *Limnol. Oceanogr.* **1990**, *35*, 16–23. [CrossRef]
47. Thouvenot, A.; Debroas, D.; Richardot, M.; Devaux, J. Impact of natural metazooplankton assemblage on planktonic microbial communities in a newly flooded reservoir. *J. Plankton Res.* **1999**, *21*, 179–199. [CrossRef]
48. Mohr, S.; Adrian, R. Reproductive success of the rotifer *Brachionus calyciflorus* feeding on ciliates and flagellates of different trophic modes. *Freshwater Biol.* **2002**, *47*, 1832–1839. [CrossRef]
49. Devetter, M.; Sed'a, J. Rotifer fecundity in relation to components of microbial food web in a eutrophic reservoir. *Hydrobiologia* **2003**, *504*, 167–175. [CrossRef]
50. Zhan, A.; Hulak, M.; Sylvester, F.; Huang, X.; Adebayo, A.A.; Abbott, C.L.; Adamowicz, S.J.; Heath, D.D. High sensitivity of 454 pyrosequencing for detection of rare species in aquatic communities. *Methods Ecol. Evol.* **2013**, *4*, 558–565. [CrossRef]
51. Zaiko, A.; Samuiloviene, A.; Ardura, A.; Garcia-Vazquez, E. Metabarcoding approach for nonindigenous species surveillance in marine coastal waters. *Mar. Pollut. Bull.* **2015**, *100*, 53–59. [CrossRef]
52. Comtet, T.; Sandionigi, A.; Viard, F.; Casiraghi, M. DNA (meta) barcoding of biological invasions: A powerful tool to elucidate invasion processes and help managing aliens. *Biol. Invasions* **2015**, *17*, 905–922. [CrossRef]
53. Kelly, R.P.; Port, J.A.; Yamahara, K.M.; Martone, R.G.; Lowell, N.; Thomsen, P.F.; Mach, M.E.; Bennett, M. Environmental monitoring. Harnessing DNA to improve environmental management. *Science* **2014**, *344*, 1455–1456. [CrossRef]

 *applied sciences*

Article

Comparison of Water Sampling between Environmental DNA Metabarcoding and Conventional Microscopic Identification: A Case Study in Gwangyang Bay, South Korea

Dong-Kyun Kim [1], Kiyun Park [1], Hyunbin Jo [1] and Ihn-Sil Kwak [1,2,*]

[1] Fisheries Science Institute, Chonnam National University, Yeosu 59626, Korea
[2] Faculty of Marine Technology, Chonnam National University, Yeosu 59626, Korea
* Correspondence: iskwak@chonnam.ac.kr; Tel.: +82-61-659-7148

Received: 2 July 2019; Accepted: 7 August 2019; Published: 9 August 2019

Abstract: Our study focuses on methodological comparison of plankton community composition in relation to ecological monitoring and assessment with data sampling. Recently, along with the advancement of monitoring techniques, metabarcoding has been widely used in the context of environmental DNA (eDNA). We examine the applicability of eDNA metabarcoding for effective monitoring and assessment of community composition, compared with conventional observation using microscopic identification in a coastal ecosystem, Gwangynag Bay in South Korea. Our analysis is based primarily on two surveys at a total of 15 study sites in early and late summer (June and September) of the year 2018. The results of our study demonstrate the similarity and dissimilarity of biological communities in composition, richness and diversity between eDNA metabarcoding and conventional microscopic identification. It is found that, overall, eDNA metabarcoding appears to provide a wider variety of species composition, while conventional microscopic identification depicts more distinct plankton communities in sites. Finally, we suggest that eDNA metabarcoding is a practically useful method and can be potentially considered as a valuable alternative for biological monitoring and diversity assessments.

Keywords: coastal ecosystem; eDNA; metabarcoding; microscopy; monitoring and assessment

---

1. Introduction

Environmental DNA (eDNA) is defined as genetic material indirectly obtained from a wide variety of environmental samples (e.g., air, water, and soil), rather than directly sampled from macro- and micro-organisms [1]. Since a specific region of DNA sequences accommodates the information about the identification of specific organisms of interest, eDNA collected from an environmental sample encompasses a variety of species information in an ecosystem [2]. The idea of eDNA was initiated from extracting the nucleic acids of microbes directly from environmental samples [1,3–5].

Nowadays, DNA across diverse taxonomic groups has been widely searched in the context of genome projects [6,7]. The rapid advancement of molecular technology, such as amplification using polymerase chain reaction (PCR), facilitates applications of DNA-based approaches that highlight the capacity of analysis to detect a variety of macro- and micro-organisms within the same sample. DNA-based identification has been regarded as efficient alternatives in terms of both time and cost in ecological research [8,9]. This analytical technique can be applied either to a single species/taxon using specific primers or to multiple species/taxa using generic primers in accordance with research objectives. DNA metabarcoding is a rapid method for assessing biodiversity from environmental bulk samples. In particular, rapidly growing next-generation sequencing (NGS) techniques have recently allowed comprehensive surveys for biological monitoring and assessment [8,10]. To this end, a growing body of literature has put special emphasis on the advantages of metabarcoding, highlighting its usefulness for ecological management [2,9,11–15]. Accordingly, a new type of DNA-based identification method has been developed as DNA metabarcoding, and widely introduced with plenty of applicable potentials for biological monitoring and assessment [16–18]. Specifically, eDNA metabarcoding has been newly proposed to assess the status (e.g., healthy, threatened, or degraded) of an ecosystem by detecting single (rare) and/or multiple (abundant) species in terms of biodiversity [12,13,19]. Despite the relatively short history, eDNA metabarcoding is appealing for monitoring and assessment of ecosystems due to its species detectability, cost and effort efficiency, and no environmental disturbance [18].

In coastal marine ecosystems, plankton communities play a pivotal role in food chain flow and biogeochemical cycles [20]. Particularly, zooplankton communities including both mero- and holo-zooplankton exert large influences on fish biomass and fisheries resources especially associated with juvenile growth [21]. Conventional microscopic identification (CMI) methods have mostly been used to estimate the richness and abundance of plankton communities in an aquatic ecosystem [22,23]. CMI might be limited in taxonomic identification, because the resultant data quality depends upon expertise and subjectivity of the scientists, and may cause disturbance to the habitat, and it is difficult to detect rare and endangered species [2,24]. In contrast, an eDNA analysis contains competitive advantages over CMI in detecting rare or invasive species [25]. In addition, given the high cost and large efforts for data collection and analysis in CMI, eDNA metabarcoding sheds light on efficient monitoring and assessment of a target ecosystem [8,18]. Furthermore, the rapid biological responses/changes to ambient physicochemical conditions lead to high demands on a new method that is fast and inexpressive, such as NGS-based metabarcoding [8]. Yet, the applications of eDNA have not been covered as widely as we wished, because of its short history, and to date have focused more on paleoecology and endangered species [13,19].

In the sense that the eDNA metabarcoding is highly appealing for finding cryptic aquatic species in biological monitoring and assessment, our study focuses on testing the potential of eDNA metabarcoding in order to monitor coastal plankton communities and assess biodiversity in comparison to CMI. Hence, the aim of our study is to identify spatial and temporal heterogeneity of plankton community dynamics in Gwangyang Bay of South Korea, characterizing predominant species and ambient water quality conditions. Finally, we discuss the potential values of eDNA metabarcoding as an alternative approach for ecological monitoring and rapid assessment.

2. Materials and Methods

2.1. Description of the Study Site

Gwangyang Bay is located in the south coast of Korean peninsula (Figure 1). In terms of morphological features of the bay, water depth varies from 10 m at the Seomjin River estuary to 50 m at the outer bay. The bay has a semi-diurnal tidal cycle. The bay receives a large discharge (ca. annually 2298 mega MT year^{-1}, equivalent to 72.8 m^3 s^{-1}) from Seomjin River [26]. It appears that a significant amount of nutrients (19.7 × 10^3 moles N day^{-1}, 0.1 × 10^3 moles P day^{-1}, 18.2 × 10^3 moles Si day^{-1} in average) come to the bay from the Seomjin River catchment (ca. 5000 km^2) [27]. Since the Seomjin River estuary relative to the Korean river estuaries remains open without barrages, the water mass between river and ocean exchanges more actively. This dynamic condition of the bay tends to shape great primary productivity and high biological diversity. From both an ecological and economical points of view, Gwangyang Bay (ca. 450 km^2 from the estuary to the outer bay) is the most productive coastal area in Korea. Specifically, Jeonnam Province containing Gwangyang Bay comprised 71% (1,297,815 MT year^{-1}) of aqua-cultural resources in a national scale as of 2016 (KOSIS, [28]). In addition, a large industrial area (e.g., oil refineries and steel plants) near the bay can be regarded as a significant pollution source. Thus, the intermittent release of various pollutants might be another factor disturbing water quality and benthic sediments [29].

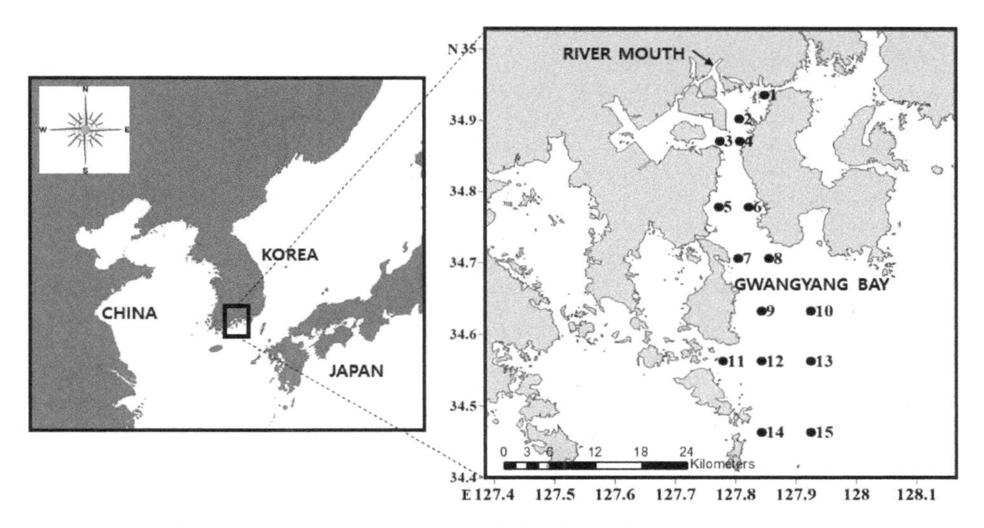

Figure 1. Map of the study sites (black closed circles) in Gwangyang Bay.

2.2. Sampling and Data Collection

The survey was conducted in June and September 2018, respectively. The total number of sampling sites was fifteen, and encompassed the extensive area from the Seomjin River estuary to the outer Gwangyang Bay (Figure 1). The water samples were collected vertically from sediments to surface (depth: 10–50 m). For marine plankton sampling, a 200 μm mesh-sized net was used. The corresponding water volume (ca. 7560 L; 7.56 m^3) was calculated by a flow-meter equipped in front of the net inlet. Zooplankton samples were identified and counted under a dissecting microscope (SV11, Zeiss and SZ60, Olympus, Tokyo, Japan), according to Chihara and Murano [30]. Water temperature and salinity were measured on site using a portable probe (Professional Plus, YSI, Yellow Springs, OH, USA). Nutrient and chlorophyll *a* concentrations (Chl-*a*) were analyzed in the lab using the collected water samples. Specifically for the measurement of phosphorus, nitrogen, and Chl-*a*, automatic water quality analyzer (AutoAnalyzer 3 HR, Seal Analytical Inc., Mequon, WI, USA) was used, and we adapted the standard analytical methods proposed by the Korea Ministry of Oceans and Fisheries (downloadable from http://www.mof.go.kr/jfile/readDownloadFile.do?fileId=MOF_ARTICLE_5689&fileSeq=1). For Chl-*a* measurement and eDNA metabarcoding, the water samples (1 L per sample) were immediately filtered in the lab, using a 0.45 μm pore-size membrane (MFS membrane filter, Advantec, Irvine, CA, USA). The membrane for Chl-*a* was then, homogenized after acetone extraction prior to the spectrophotometry. The membrane for eDNA was preserved at −80 °C. Organic and inorganic carbon concentrations were measured using a carbon analyzer (vario TOC cub, Elemetar, Langenselbold, Germany) on the basis of 850 °C combustion catalytic oxidation methods.

2.3. DNA Extraction and Metagenomic Sequencing

Genomic DNA was extracted by means of PowerSoil® DNA Isolation Kit (Cat. No. 12888, MO BIO, Germantown, MD, USA) in compliance with the manufacturers' protocol. Extracted DNA for sequencing was prepared according to the Illumina 18S Metagenomic Sequencing Library protocols (San Diego, CA, USA). DNA quantity, quality, and integrity were measured by PicoGreen (Thermo Fisher Scientific, Waltham, MA, USA) and VICTOR Nivo Multimode Microplate Readers (PerkinElmer, Akron, OH, USA). For our study, the 18S rDNA V9 barcode was used, because it has often been applied to semi-quantitatively estimate relative abundances within a sample [31–33]. More specifically, we obtained the primer information from a study by Guo et al. [33], which also followed the universal primers for 18S V9 region designed by Amaral-Zettler et al. [34]. The primer sequences are as follows: 18S V9 primer including adaptor sequence (Forward Primer: 5′ TCGTCGGCAGCGTCAGATGTGT ATAAGAGACAG**CCCTGCCHTTTGTACACAC** 3′, Reverse Primer: 5′ GTCTCGTGGGCTCGGA GATGTGTATAAGAGACAG**CCTTCYGCAGGTTCACCTAC** 3′, the primers are in bold). The PCR master mixture of 25 μL (Macrogen Inc., Seoul, Korea) comprised 2 μL of genomic DNA (1 ng/μL), 1.25 μL of each primer (5 μM), 5 μL of 5 × Herculase II Reaction Buffer, 0.25 μL of dNTP mix (100 mM), 0.5 μL of Herculase II Fusion DNA polymerase (Agilent, Waldbronn, Germany), and 14.75 μL of PCR Grade water. To amplify the target region attached with adapters, as a first PCR process, the extracted DNA was amplified by 18S V9 primers with one cycle of 3 min at 95 °C, 25 cycles of 30 s at 95 °C, 30 s at 55 °C, 30 s at 72 °C, and a final step of 5 min at 72 °C for amplicon PCR product. As a second process, to produce indexing PCR, the first PCR product was subsequently amplified with one cycle of 3 min at 95 °C, 8 cycles of 30 s at 95 °C, 30 s at 55 °C, 30 s at 72 °C, and a final step of 5 min at 72 °C. A subsequent limited-cycle amplification step was performed to add multiplexing indices and Illumina sequencing adapters (Figure 2). The final products were normalized and pooled using the PicoGreen (ThermoFisher Scientific, Waltham, MA, USA), and the size of the libraries was verified using the LabChip GX HT DNA High Sensitivity Kit (PerkinElmer, Waltham, MA, USA).

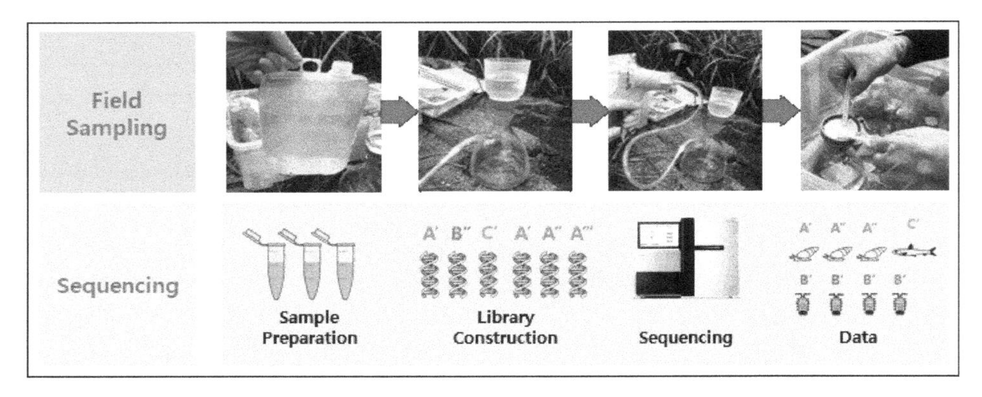

Figure 2. Analytical procedure of environmental DNA (eDNA) extraction and metagenomic sequencing.

A sequencing library is prepared by random fragmentation of the DNA or cDNA sample, followed by 5′ and 3′ adapter ligation. Alternatively, "tagmentation" combines the fragmentation and ligation reactions into a single step that greatly increases the efficiency of the library preparation process. Adapter-ligated fragments are then PCR amplified and gel purified. The PCR products were sequenced using the MiSeq™ platform (Illumina, San Diego, CA, USA) from commercial service (Macrogen Inc., Republic of Korea). In total, filtered 6,151,975 paired-end reads from the 30 samples were generated on the platform, of which 97.11% passed Q30 (Phred quality score > 30) in this study. Raw reads were trimmed with CD-HIT-OTU and chimeras were identified and removed using rDNATools. For paired-end merging, FLASH (Fast Length Adjustment of Short reads) version 1.2.11 was used. Each sample yielded paired-end reads ranging from 21,101–299,305 reads (mean: 180,940 reads), and all samples exhibited the saturation of the number of operational taxonomic units (OTUs) by rarefaction curve analysis (see Appendix A). Merged reads were processed and were clustered into OTUs using a bioinformatic algorithm, UCLUST [35], at a 97% OTU cutoff value (352 OTUs in gamma-diversity). The resulting 552 OTUs were classified into 19 genus-level taxonomic groups (those representing < 0.04% abundance were not plotted). Taxonomy was assigned to the obtained representative sequences with BLAST (Reference DB: NCBI—18S) [36] using UCLUST [35]. For the aforementioned processes of BLAST and UCLUST, we used an open-source bioinformatics pipeline for performing microbiome analysis, QIIME version 2 [37].

2.4. Analytical Methods

The self-organizing map (SOM) is an unsupervised neural network as machine learning, and it is commonly known as a powerful tool for pattern recognition from complex data [38]. In ecological research, the SOM has recently been considered as a more appropriate multivariate analysis than other conventional statistical approaches [39]. The SOM is robust and suitable in providing comprehensive views on highly complex and multi-dimensional data through reducing the data dimension. The efficiency of SOMs in information extraction was demonstrated across different hierarchical levels of life from molecules to ecosystems [40]. Several studies showed that the SOM was robust enough to capture the nonlinear pattern of an ecosystem [39,41,42]. For these reasons, the SOM has been extensively applied to pattern recognition in various ecological domains including benthic macroinvertebrates [43,44], plankton communities [45–48], dissolved organic matters [49], fish assemblages [50,51], and biomanipulation assessment [52,53].

In the SOM analysis, a total of 33 variables were used including six physicochemical parameters, 27 dominant plankton populations (10 from the eDNA, and 17 from the CMI samples). In selecting the number of variables, we only included the plankton communities, of which abundance was greater than 5% of the total abundance. That is, otherwise, the variables would contain too many zero values which could lead to topological biases in the SOM visualization. The SOM size was determined by the rule of $5\sqrt{\text{sample size}}$ [54]. The SOM model was developed using MATLAB 6.1 (The MathWorks Inc., Natick, MA, USA) and the SOM Toolbox (Helsinki University of Technology, Espoo, Finland).

For assessment of richness and diversity, the former simply equals to number of species, while the latter is based on the Shannon–Weaver index ($H' = -\sum p_i \ln p_i$, p_i indicates a fraction of ith species) [55]. In calculating those biological indices, we excluded the taxonomical groups from eDNA samples, such as bacteria, mammals, reptiles, terrestrial plants, and amphibians, because the comparison between two different methods should be done at the same level of analytical resolution.

3. Results

3.1. Comparative Estimation of Coastal Biota between eDNA Metabarcoding and CMI

A variety of plankton communities were observed through the two identification methods in Gwangyang Bay. There were differences in the number of identified communities between the two methods (Table 1). In terms of quantity, eDNA metabarcoding seemed to be capable of detecting more species. The average numbers of observed (identified and unidentified) species from the eDNA samples were 27.9 (min to max: 20–36) in June and 49.8 (min to max: 13–72) in September, while those from the CMI were 19.6 (min to max: 12–23) and 18.9 (min to max: 12–24), in June and September, respectively (Table 1). Albeit comparing only with the identified species, we found that the number of species was higher in the eDNA samples than the CMI. On the other hand, in terms of the capability of identification of the eDNA metabarcoding, the unidentified species groups comprised 38% in June and 19% in September (Table 1). Accordingly, in Gwangyang Bay, the eDNA samples identified more species in a higher proportion in September.

In the eDNA samples, the richness values in September were as twice high as those in June (Table 1). The number of identified species was lower in June (mean ± S.D.: 20.2 ± 3.3) than in September (41.7 ± 15.6), which was quite consistent across the study sites. In addition, this pattern was similarly observed from diversity values of the eDNA samples (averages: 1.0 in June and 2.0 in September). The spatial variation of the richness was also lower in June (coefficient of variation: ca. 15%) than in September (ca. 40%). Namely, the heterogeneity of plankton distribution became large in late summer. On the contrary, in the CMI samples, the species richness did not differ between early and late summer; the values of mean and S.D. were 19.6 ± 3.1 in June, and 18.9 ± 3.5 in September. Notably, the level of diversity was comparatively higher in June (2.3 ± 0.2) than in September (1.6 ± 0.2), which was counter to the diversity pattern from the eDNA samples. Considered as a whole, the temporal changes of biological communities seem to be more distinct, compared to their spatial variation. Nonetheless, we also note some discrepancy of the results in diversity between the two identification methods.

Table 1. Richness and Shannon diversity of the samples between water eDNA and conventional microscopic identification (CMI) in Gwangyang Bay. The numbers in the brackets indicate the number of unidentified groups.

	June				September			
	eDNA		CMI		eDNA		CMI	
Site	Richness	Diversity	Richness	Diversity	Richness	Diversity	Richness	Diversity
GY1	28 (6)	1.01	23	2.37	54 (9)	1.99	20	1.62
GY2	28 (7)	1.65	23	2.43	59 (10)	2.39	17	1.54
GY3	20 (5)	1.08	23	2.55	34 (5)	1.50	12	1.34
GY4	23 (6)	1.35	22	2.46	72 (11)	2.66	16	1.70
GY5	25 (7)	1.18	22	2.36	44 (7)	1.65	13	1.35
GY6	22 (6)	0.72	20	2.33	62 (8)	2.13	19	1.55
GY7	29 (8)	0.87	12	1.96	13 (3)	0.24	18	1.78
GY8	33 (10)	1.08	18	2.24	29 (6)	1.91	23	1.90
GY9	35 (11)	1.59	22	2.48	58 (8)	2.31	24	1.61
GY10	27 (8)	1.05	20	2.37	48 (10)	2.44	18	1.41
GY11	36 (11)	1.50	16	1.92	72 (11)	2.69	18	1.72
GY12	34 (10)	0.56	18	2.31	35 (4)	2.02	20	1.61
GY13	31 (8)	0.73	18	2.38	46 (8)	1.57	23	1.81
GY14	24 (6)	0.64	18	1.75	45 (9)	1.96	22	1.99
GY15	24 (7)	0.70	19	2.05	64 (12)	2.20	20	1.66
Mean	27.9 (7.7)	1.0	19.6	2.3	49.7 (8.1)	2.0	18.9	1.6
S.D.	5.0 (1.9)	0.4	3.1	0.2	17.8 (2.7)	0.6	3.5	0.2

To evaluate the consistency of detection and identification of marine plankton groups, we compared the differences of community composition between eDNA and CMI samples (Figure 3a,b). Although various groups were detected by eDNA metabarcoding, the community composition was based on the identified groups in eDNA samples, in comparison with those from the CMI samples. In the higher rank of taxonomical classification (>phylum), the eDNA samples comprised 28% of phytoplankton (i.e., algae) and 15% of zooplankton (i.e., Copepoda) (Figure 3a), whereas the CMI samples showed 64% of zooplankton (Figure 3b).

In the eDNA samples, the dominant groups in phytoplankton were diatoms (e.g., *Thalassiosira* spp.) and dinoflagellates (*Hematodinium* spp.). In zooplankton, the dominant groups were marine calanoid copepods such as *Acartia* spp. and *Centropages* spp. in Gwangyang Bay. Crustacea occupied 17% of the identified species, and were primarily comprised of Amphipoda (e.g., *Caprella* spp.), Cirripedia (e.g., barnacles), and Decapoda (e.g., *Corophium* spp.). Cnidaria and Mollusca also engaged species richness of 24% in our study area (Figure 3a). The former consisted mainly of small polyp stony coral, such as *Acropora* spp., and the latter mostly comprised bivalves, such as *Crassostrea* spp. and *Musculista* spp. In addition, several groups, which were relatively low proportionally in CMI, were also well identified, including Annelida (5%), Chaetognatha (4%), Echinoderma (3%), and fish (4%). Particularly for fish, the identification of fish species was quite limited in the eDNA samples, and hence only three genera were identified (*Arnoglossus*, *Engraulis*, and *Oryzias* spp.).

By comparison, the CMI samples showed different proportion in species richness (Figure 3b). The main composition (64%) of zooplankton comprised Cladocera (e.g., *Evadne* spp. and *Podon leuckarti*) as well as Copepoda (e.g., 15 calanoid species and three cyclopoid species). Conversely, a limited number of phytoplankton was identified in the CMI samples, compared to the eDNA samples. The identified phytoplankton were mostly dinoflagellates which were mainly *Noctiluca scintillans*. Crustecea occupied 11% of species richness. Similar to the identified species from the eDNA samples, they were primarily composed of Amphipoda, Cirripedia, and Decapoda. However, most of them were in forms of larvae which was unable to be identified specifically in the CMI samples. Other specific groups were observed in a small proportion (3%: Annelida, Chaetognatha, Cnidaria, and Echinoderma, and 5%: Fish and Mollusca, see Figure 3b). Nevertheless, in a finer resolution, there was some commonality of species groups between eDNA and CMI samples (Table 2). In both samples, several genera including *Acartia*, *Acropora* and *Centropages*, were commonly observed. At the Gwangyang Bay, *Acartia* spp. were commonly predominant in early summer, while *Centropages* spp. were relatively predominant in late summer. *Acropora* spp. were primarily observed from the eDNA samples around the inner bay in early summer. At the outer bay, including at site 14 and site 15, a dinoflagellate group of *Hematodinium* was relatively abundant, especially in the eDNA samples. In contrast, *Hematodinium* was not detected by CMI in the same area. Moreover, *Oithona* spp. were most predominant in this area, but were relatively less abundant in the outer bay, compared to the inner bay.

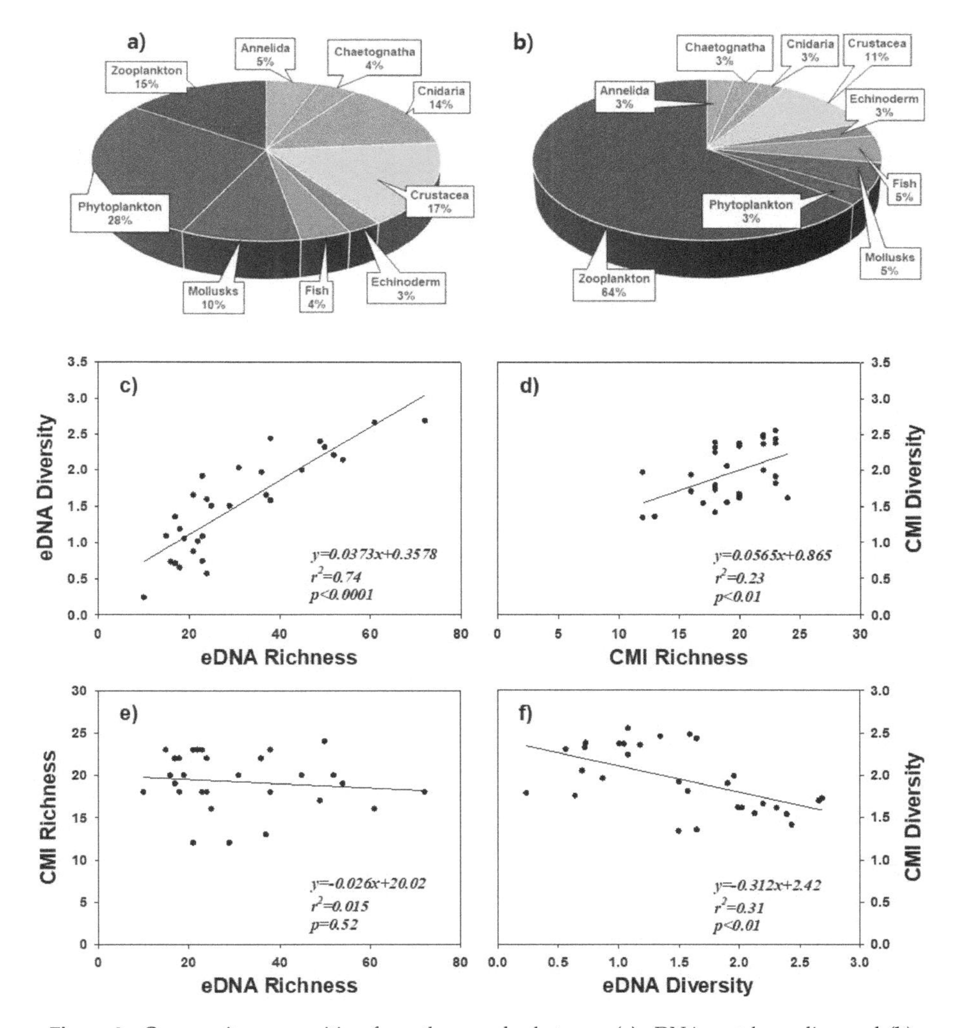

Figure 3. Community composition from the samples between (**a**) eDNA metabarcoding and (**b**) conventional microscopy identification (CMI). The scatter plots indicate the relationships between species richness and Shannon–Weaver diversity on (**c**) eDNA, (**d**) CMI, and those (**e**) between richness and (**f**) between diversity of the two methods, respectively.

Table 2. Dominant plankton groups observed during the summer season (June and September) in Gwangyang Bay.

Site	eDNA Metabarcoding	CMI
GY1	*Acropora, Candacia, Caprella, Oryzias*	*Acartia, Paracalanus*
GY2	*Acropora, Candacia, Caprella, Corophium, Oryzias*	*Acartia, Corycaeus, Centropages, Corycaeus, Oithona, Paracalanus, Sagitta*
GY3	*Acartia, Centropages*	*Acartia, Noctiluca, Oithona, Paracalanus, Sagitta*
GY4	*Acartia, Acropora, Caprella, Corophium*	*Acartia, Corycaeus, Noctiluca, Oithona, Paracalanus, Sagitta*
GY5	*Acartia, Acropora, Centropages*	*Acartia, Noctiluca, Paracalanus, Sagitta*
GY6	*Acropora, Hematodinium*	*Acartia, Corycaeus, Noctiluca, Oithona, Paracalanus, Sagitta*
GY7	*Acartia*	*Centropages*
GY8	*Acropora, Caprella,*	*Centropages, Noctiluca*
GY9	*Acartia, Acropora, Candacia, Centropages, Hematodinium*	*Centropages, Noctiluca*
GY10	*Acropora, Thalassiosira*	*Centropages, Corycaeus, Sagitta*
GY11	*Candacia, Caprella, Centropages*	*Centropages*
GY12	*Centropages, Hematodinium,*	*Centropages, Paracalanus*
GY13	*Candacia, Centropages*	*Centropages, Paracalanus*
GY14	*Hematodinium*	*Oithona*
GY15	*Hematodinium*	*Oithona*

3.2. Relationships of Biotic Information between eDNA and CMI Samples

To examine consistency of biological information between different sampling strategies, the relationships between species richness and diversity were comparatively assessed. In both eDNA and CMI samples, species richness and diversity were positively correlated with each other (Figure 3c,d). The eDNA samples showed stronger signal of the positive relationship between species richness and diversity than the CMI samples, and the interpretability of species richness on corresponding diversity was three times higher in the eDNA samples (r^2 = 0.74) than in the CMI samples (r^2 = 0.23). Although both samples showed the significant relationships between the two, the relationship was clearer in the eDNA samples. On the other hand, we also examined the relationships between the richness values and between the diversity values (Figure 3e,f). There was no statistical significance between the richness values (i.e., eDNA versus CMI samples) (Figure 3e). In addition, although the diversity values exhibited statistical significance in their relationship, the signal was slightly negative, which was counterintuitive (Figure 3f). In consequence, it appeared that the information obtained from the same methodology was consistent enough to project the relationship between species richness and diversity. Conversely, it was found that there was a discrepancy of biotic information between eDNA and CMI samples.

3.3. Assessment of Biogeochemical Characteristics in Gwangyang Bay

The clustering analysis using the SOM characterized biogeochemical features of Gwangyang Bay into four distinct patterns. The four clusters determined by the SOM shaped spatiotemporal heterogeneity of the data samples at Gwangyang Bay (Figure 4 and Appendix B). It is remarkable to discern the spatiotemporal pattern that cluster 1 included site 1 to site 6 of June, cluster 2 site 7 to site 15 of June, cluster 3 site 1 to site 8 of September, and cluster 4 site 9 to site 15 of September as well as site 14 and site 15 of June (Figure 4a). In addition, the estimate of neighboring distances among the clusters indicated that the clusters were firstly separated as top (cluster 3 and cluster 4) and bottom (cluster 1 and cluster 2). As a consequence, the clustering result manifested that plankton community of Gwangyang Bay was primarily characterized by seasonal influences between early and late summer (i.e., June and September at Gwangyang Bay), and then was spatially distinguished. Strictly speaking, site 14 and site 15 of June were grouped as cluster 4 which represented the outer bay of late summer, but they were placed on the bottom of cluster 4, which was characterized as the outer bay of early summer. Namely, these two sites appear to represent similar features on coastal plankton community, regardless of temporal changes in summer.

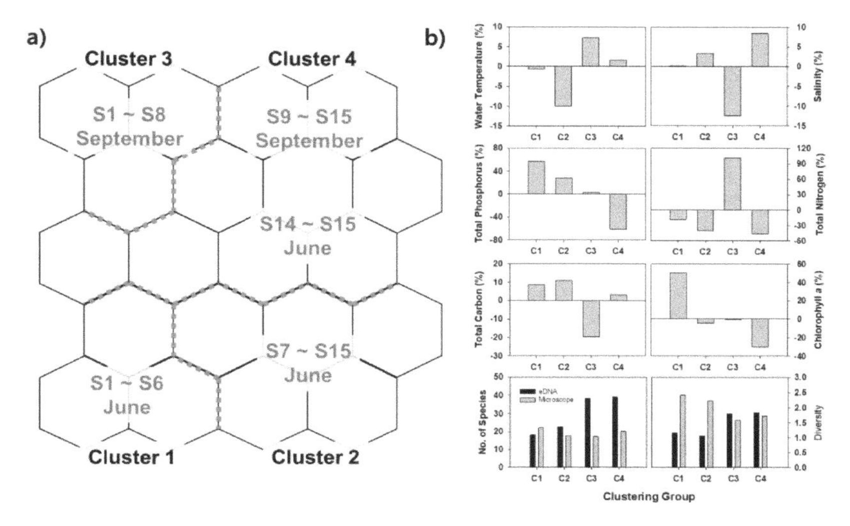

Figure 4. Clustering result (**a**) of the data of water eDNA and CMI based on the self-organizing map. The right panels (**b**) present the corresponding physical, chemical, and biological conditions in Gwangyang Bay. The horizontal lines of zero indicate corresponding grand average values (water temperature: 25.6 °C, salinity: 29.3 psu, TP: 0.049 mg L^{-1}, TN: 0.45 mg L^{-1}, TC: 22.4 mg L^{-1}, Chl-*a*: 4.36 mg L^{-1}).

Several water quality parameters delineated ambient physicochemical conditions associated with plankton community in Gwangyang Bay (Figure 4b). Water temperature was relatively lower in cluster 2 and higher in cluster 3 among the four groups. The higher salinity of the outer bay matched well with cluster 2 and cluster 4. Cluster 3 represented the inner bay of the summer, exhibiting lower salinity was higher water temperature. Concerning nutrient concentration, total phosphorus (TP) concentrations were higher in June (cluster 1 and cluster 2 in Figure 4) than in September (cluster 3 and cluster 4 in Figure 4). In the spatial scale, TP was higher at the inner bay (cluster 1 and cluster 3 in Figure 3) than at the outer bay (cluster 2 and cluster 4 in Figure 4). In addition to TP, total nitrogen (TN) concentrations were conspicuously high in cluster 3, which represented the inner bay in late summer. Total carbon (TC) concentrations displayed opposite patterns against TN. Among the four clustering groups, chlorophyll *a* (Chl-*a*) concentrations were highest at the inner bay in early summer, while were lowest at the outer bay in late summer. In view of biotic information, the number of species was relatively higher in cluster 3 and cluster 4 (September) based on the eDNA samples, while the diversity indices were comparatively higher in cluster 1 and cluster 2 (June) based on the CMI samples (Figure 4).

4. Discussion

4.1. Congruence of Taxonomic Information between eDNA Metabarcoding and CMI

Many of recent studies have strived to profile and quantify taxonomic composition of plankton communities using either eDNA metabarcoding or CMI [14,24,56]. Among them, a few studies have reported a degree of disagreement between the two pronged identification methods [32,57]. In this respect, our study also presented some disagreement, between eDNA and CMI samples, in community composition (Figure 3a,b as well as in relationships of biotic information (Figure 3e,f). Some pieces of literature on eDNA monitoring have enumerated possible reasons to explain the discrepancy between the two identification methods. It is reported that the capacity of identification between molecular and morphological datasets could have mainly caused the disagreement [24,58]. That is, specimen identification can vary along accuracy of molecular reference databases [59]. Therefore, the establishment of well-curated databases of reference DNA sequences for identified specimens is essential in the field of eDNA metabarcoding to make the taxonomic information congruent with CMI. Additionally, there is another concern with the drawback of eDNA metabarcoding associated with technical biases/difficulties, such as copy number variation in the process of polymerase chain reaction (PCR) [60]. Related to a primer, its amplification and binding affinity are critical factors to bring about taxonomic biases in eDNA detection [61–65]. In terms of sensitivity of species detection, CMI-based assessment is also subject to an unpredictable, but probably significant, bias due to the presence of cryptic species [66]. Particularly in our study, marine calanoid copepods, *Candacia*, were only detected by eDNA metabarcoding in a very low proportion of <5%. However, we also admit that taxonomic misclassification due to lack of expertise and difficult to impossible taxonomic determination rather than just cryptic species also causes bias.

With these concerns in mind, our results on the community composition might be influenced by the primer amplification effects (Figure 3a,b). The previous related research reported some technical biases against low-abundant taxa in delineating microbial diversity [63]. In fact, while Cnidaria comprised 3% in CMI, they were 14% in eDNA samples. Likewise, Mollusks occupied 5% in CMI, but did 10% in eDNA samples (Figure 3a,b). In contrast to these differences, the compositional changes between the two samples were not significant for the rest low-abundant taxa containing Annelida, Chaetognatha, Echinoderm, and fish (Figure 3a,b). Namely, our results showed that low-abundant taxa could always be overestimated in eDNA metabarcoding. These results of difference and variation might be associated with several reasons. Firstly, eDNA metabarcoding is highly sensitive to detecting species. This high sensitivity is advantageous in identifying low-abundant/cryptic species. However, it can also lead to variations originating not only from organisms that are a few miles away from the sampling site but also from food items hidden in organisms. In addition, abundance estimates are possibly erroneous because many small organisms could generate the same number of sequence reads as a few large organisms. Secondly, although it is relatively unexplored, the copy number variation derived from the technical bias during the PCR process is another factor leading to inaccurate estimation [60,63]. Lastly, CMI is also error-prone depending on expertise/experience and specimen size. Therefore, we notice that eDNA may not be able to fully present diversity yet.

Despite some discrepancy between the eDNA and CMI samples, one highlighting point is the relational consistency in richness and diversity. Traditionally, plankton community assessment on richness and diversity has been complicated and time-consuming. However, compared to CMI, the eDNA metabarcoding also presented a positive relationship between richness and diversity (Figure 3c,d). Furthermore, while CMI exhibited a shorter range of richness and diversity (Figure 3d), the eDNA metabarcoding displayed a wider range (Figure 3c). Although its accuracy is another issue as previously mentioned, therefore, our study explicitly accounts for better capability of detection and identification by means of metabarcoding skills.

4.2. Potential Values of an eDNA Approach for Biological Monitoring and Assessment

Most conventional approaches for biological monitoring and assessment were based primarily on microscopy. Due to the time consumption and expertise requirement for identification in species level, the current environmental monitoring and assessment of community composition highly demand new alternative technologies in terms of cost efficiency. In this regard, eDNA metabarcoding has been deemed as a promising tool for species detection and identification [58]. Particularly in plankton research, the eDNA approach helped reveal a previously hidden taxonomic richness for diverse meroplankton, such as Bivalvia, Gastropoda, and Polychaeta, which are relatively hard to identify in CMI [67]. Our study also advocates that a wider variety of species, including the aforementioned meroplankton, were detected in the eDNA samples (Table 1).

At the same time, however, we recognize that some discrepancies of abundance between metabarcoding and CMI have been contentious [62,68,69]. This discrepancy may limit the scope of eDNA research, which is also associated with the varying lengths of time to eDNA degradation in response to ambient environmental conditions [11,59,70,71]. Nevertheless, several studies have found a significant relationship between determining relative or rank abundance, highlighting the potential value of eDNA, though the variation inherent in environmental samples makes it difficult to quantify [12,32].

In our study, we found some clear patterns of coastal plankton communities in time (early vs. late summer) and space (inner vs. outer bay). From our analysis using eDNA and CMI samples, the main features of Gwangyang Bay could be characterized more clearly: (*i*) inner bay in early summer; (*ii*) outer bay in early summer; (*iii*) inner bay in late summer; and (*iv*) outer bay in late summer (Figure 5). Each characteristic was explicitly delineated by the prominent species. For example, in Gwangyang Bay, Asterozoa were predominant in early summer, *Sagitta* spp. were abundant in the inner bay, and zooplankton *Centrophages* spp. were in late summer. Dinoflagellates were separately characterized by *Noctilluca* spp. in early summer and by *Hematodinium* spp. in late summer. Although we did not use the eDNA samples solely, our spatiotemporal analysis presented the main plankton community features based on both eDNA and CMI samples. The CMI samples in addition to the eDNA make our pattern analysis more robust and reliable, because the predominant plankton would be separately presented if the eDNA and CMI samples differed significantly from each other. Thus, the information from the eDNA and CMI samples was highly similar given the subtle discrepancy of richness, diversity and their relationships (Figure 3c,d). However, we stress that the eDNA samples were good enough to delineate spatial and temporal characteristics of coastal plankton communities in Gwangyang Bay (Figure 5).

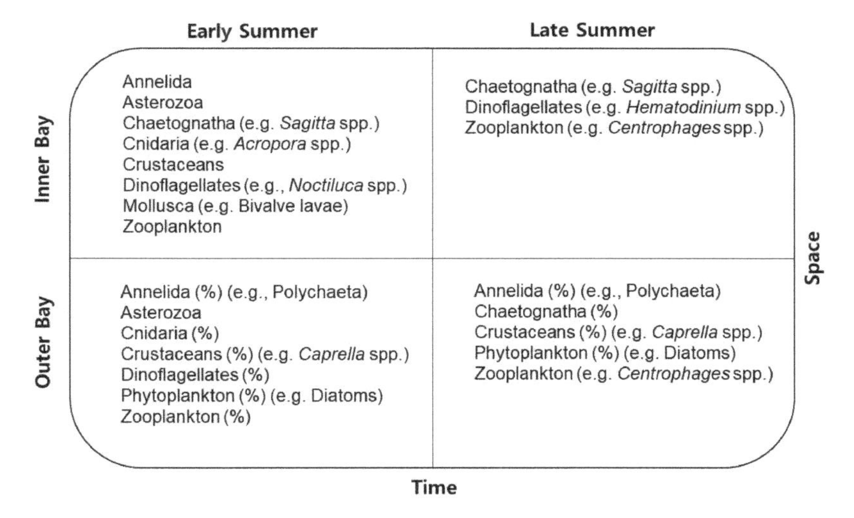

Figure 5. Main characteristics of marine plankton communities in Gwangyang Bay. The groups with % present relative abundance (derived from eDNA), and those without % present absolute abundance (derived from CMI).

In sum, we learn from our study that eDNA metabarcoding can be an effective alternative to monitor and assess entire communities from even a single sample. In addition, the eDNA metabarcoding is highly beneficial in terms of sensitivity for cryptic species and cost-efficiency for morphological identification. At the same time, however, our study also put emphasis on bio-assessment that can be affected by some information discrepancy of richness and diversity between eDNA and CMI samples. Hence, eDNA-based research should be further investigated to make the derived results become more stable. The current limited capacity of eDNA-based research is probably subject to a great deal of uncertainties associated with amplification, reference database, NGS-sequencing, and eDNA degradation [57,71]. To this end, we stress that eDNA research should be more active in order to shed light on ecosystem monitoring and assessment in future.

Author Contributions: Conceptualization: D.-K.K., K.P., H.J. and I.-S.K.; Methodology: D.-K.K. and K.P.; Formal Analysis: D.-K.K. and H.J.; Investigation: D.-K.K., K.P. and H.J.; Resources: I.-S.K.; Writing—Original Draft Preparation: D.-K.K.; Writing—Review and Editing: D.-K.K. and I.-S.K.; Supervision: I.-S.K.; Project Administration: I.-S.K.; Funding Acquisition: I.-S.K.

Funding: This research was funded by the National Research Foundation of Korea, grant number NRF-2018R1A6A1A03024314.

Conflicts of Interest: The authors declare that they have no competing interests.

Appendix A Rarefaction Curves of the 18S rDNA V9 Samples in May (A) and September (B)

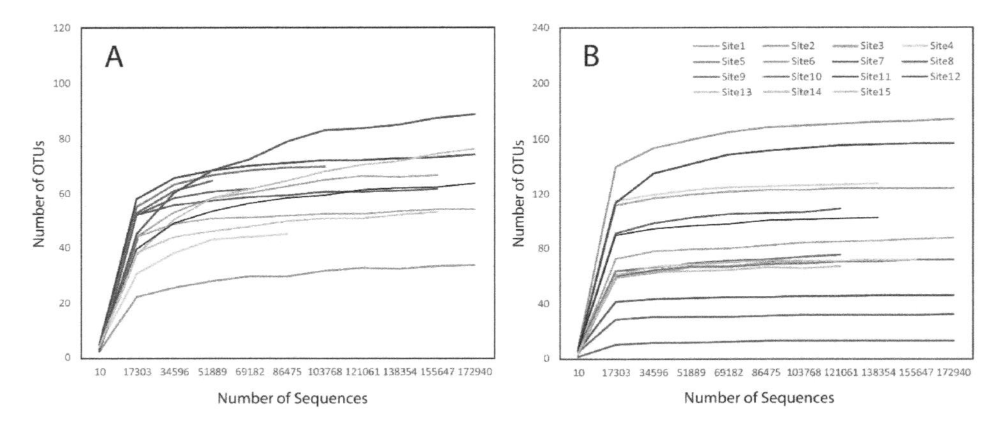

Appendix B Visualization of Explanatory Variables Derived from Self-Organizing Maps

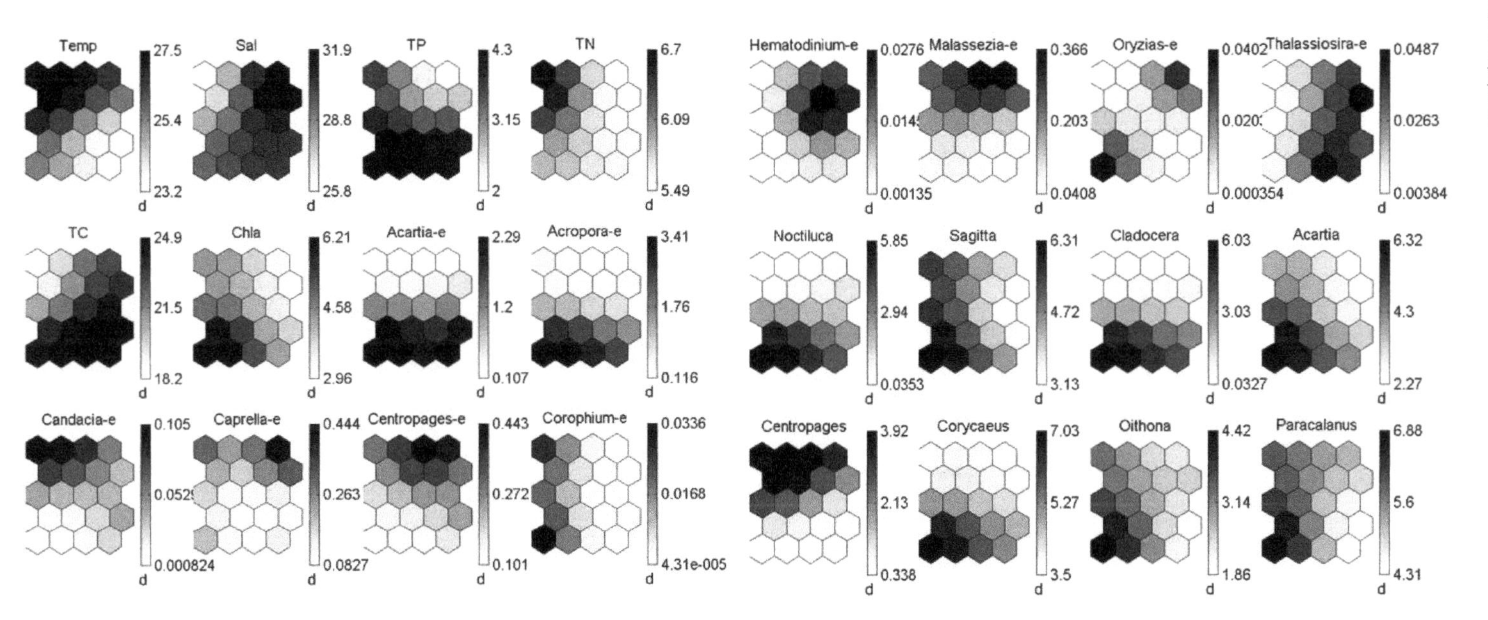

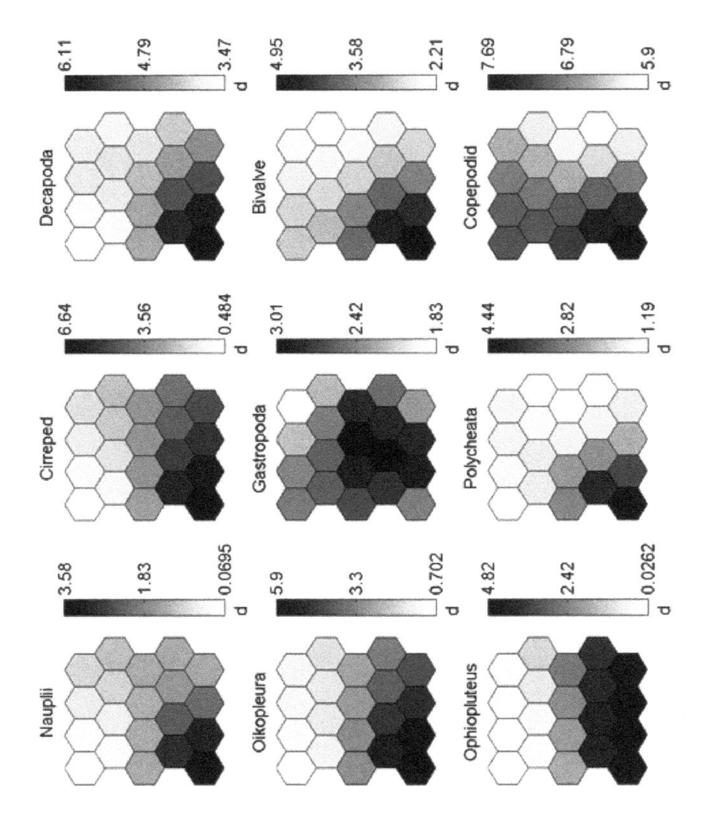
Nauplii
3.58
1.83
0.0695
d
Cirreped
6.64
3.56
0.484
d
Decapoda
6.11
4.79
3.47
d
Oikopleura
5.9
3.3
0.702
d
Gastropoda
3.01
2.42
1.83
d
Bivalve
4.95
3.58
2.21
d
Ophiopluteus
4.82
2.42
0.0262
d
Polycheata
4.44
2.82
1.19
d
Copepodid
7.69
6.79
5.9
d

References

1. Taberlet, P.; Coissac, E.; Hajibabaei, M.; Rieseberg, L.H. Environmental DNA. *Mol. Ecol.* **2012**, *21*, 1789–1793. [CrossRef] [PubMed]
2. Thomsen, P.F.; Willerslev, E. Environmental DNA—An emerging tool in conservation for monitoring past and present biodiversity. *Biol. Conserv.* **2015**, *183*, 4–18. [CrossRef]
3. Ogram, A.; Sayler, G.S.; Barkay, T. The extraction and purification of microbial DNA from sediments. *J. Microbiol. Methods* **1987**, *7*, 57–66. [CrossRef]
4. Pace, N.R.; Stahl, D.A.; Lane, D.J.; Olsen, G.J. The analysis of natural microbial populations by ribosomal RNA sequences. In *Advances in Microbial Ecology*; Marshall, K.C., Ed.; Springer US: Boston, MA, USA, 1986; pp. 1–55.
5. Olsen, G.J.; Lane, D.J.; Giovannoni, S.J.; Pace, N.R.; Stahl, D.A. Microbial Ecology and Evolution: A Ribosomal RNA Approach. *Annu. Rev. Microbiol.* **1986**, *40*, 337–365. [CrossRef] [PubMed]
6. Watson, J.D. The human genome project: Past, present, and future. *Science* **1990**, *248*, 44–49. [CrossRef]
7. Venter, J.C.; Remington, K.; Heidelberg, J.F.; Halpern, A.L.; Rusch, D.; Eisen, J.A.; Wu, D.; Paulsen, I.; Nelson, K.E.; Nelson, W.; et al. Environmental Genome Shotgun Sequencing of the Sargasso Sea. *Science* **2004**, *304*, 66. [CrossRef] [PubMed]
8. Valentini, A.; Taberlet, P.; Miaud, C.; Civade, R.; Herder, J.; Thomsen, P.F.; Bellemain, E.; Besnard, A.; Coissac, E.; Boyer, F.; et al. Next-generation monitoring of aquatic biodiversity using environmental DNA metabarcoding. *Mol. Ecol.* **2016**, *25*, 929–942. [CrossRef]
9. Cristescu, M.E. From barcoding single individuals to metabarcoding biological communities: Towards an integrative approach to the study of global biodiversity. *Trends Ecol. Evol.* **2014**, *29*, 566–571. [CrossRef]
10. Shokralla, S.; Spall, J.L.; Gibson, J.F.; Hajibabaei, M. Next-generation sequencing technologies for environmental DNA research. *Mol. Ecol.* **2012**, *21*, 1794–1805. [CrossRef]
11. Taberlet, P.; Coissac, E.; Pompanon, F.; Brochmann, C.; Willerslev, E. Towards next-generation biodiversity assessment using DNA metabarcoding. *Mol. Ecol.* **2012**, *21*, 2045–2050. [CrossRef]
12. Bohmann, K.; Evans, A.; Gilbert, M.T.P.; Carvalho, G.R.; Creer, S.; Knapp, M.; Yu, D.W.; de Bruyn, M. Environmental DNA for wildlife biology and biodiversity monitoring. *Trends Ecol. Evol.* **2014**, *29*, 358–367. [CrossRef] [PubMed]
13. Sigsgaard, E.E.; Carl, H.; Møller, P.R.; Thomsen, P.F. Monitoring the near-extinct European weather loach in Denmark based on environmental DNA from water samples. *Biol. Conserv.* **2015**, *183*, 46–52. [CrossRef]
14. Pawlowski, J.; Kelly-Quinn, M.; Altermatt, F.; Apothéloz-Perret-Gentil, L.; Beja, P.; Boggero, A.; Borja, A.; Bouchez, A.; Cordier, T.; Domaizon, I.; et al. The future of biotic indices in the ecogenomic era: Integrating (e)DNA metabarcoding in biological assessment of aquatic ecosystems. *Sci. Total Environ.* **2018**, *637–638*, 1295–1310. [CrossRef] [PubMed]
15. Darling, J.A.; Mahon, A.R. From molecules to management: Adopting DNA-based methods for monitoring biological invasions in aquatic environments. *Environ. Res.* **2011**, *111*, 978–988. [CrossRef] [PubMed]
16. Baird, D.J.; Hajibabaei, M. Biomonitoring 2.0: A new paradigm in ecosystem assessment made possible by next-generation DNA sequencing. *Mol. Ecol.* **2012**, *21*, 2039–2044. [CrossRef]
17. Goldberg, C.S.; Strickler, K.M.; Pilliod, D.S. Moving environmental DNA methods from concept to practice for monitoring aquatic macroorganisms. *Biol. Conserv.* **2015**, *183*, 1–3. [CrossRef]
18. Ruppert, K.M.; Kline, R.J.; Rahman, M.S. Past, present, and future perspectives of environmental DNA (eDNA) metabarcoding: A systematic review in methods, monitoring, and applications of global eDNA. *Glob. Ecol. Conserv.* **2019**, *17*, e00547. [CrossRef]
19. Thomsen, P.F.; Kielgast, J.; Iversen, L.L.; Wiuf, C.; Rasmussen, M.; Gilbert, M.T.P.; Orlando, L.; Willerslev, E. Monitoring endangered freshwater biodiversity using environmental DNA. *Mol. Ecol.* **2012**, *21*, 2565–2573. [CrossRef]
20. Quéré, C.L.; Harrison, S.P.; Colin Prentice, I.; Buitenhuis, E.T.; Aumont, O.; Bopp, L.; Claustre, H.; Cotrim Da Cunha, L.; Geider, R.; Giraud, X.; et al. Ecosystem dynamics based on plankton functional types for global ocean biogeochemistry models. *Glob. Chang. Biol.* **2005**, *11*, 2016–2040.
21. Romare, P.; Bergman, E.; Hansson, L.-A. The impact of larval and juvenile fish on zooplankton and algal dynamics. *Limnol. Oceanogr.* **1999**, *44*, 1655–1666. [CrossRef]

22. Descy, J.-P.; Leitao, M.; Everbecq, E.; Smitz, J.S.; Deliège, J.-F. Phytoplankton of the River Loire, France: A biodiversity and modelling study. *J. Plankton Res.* **2012**, *34*, 120–135. [CrossRef]
23. Wetzel, R.G.; Likens, G.E. *Limnological Analysis*; Springer-Verlag: New York, NY, USA, 1991; p. 429.
24. Zimmermann, J.; Glöckner, G.; Jahn, R.; Enke, N.; Gemeinholzer, B. Metabarcoding vs. morphological identification to assess diatom diversity in environmental studies. *Mol. Ecol. Resour.* **2015**, *15*, 526–542. [CrossRef] [PubMed]
25. Rees, H.C.; Maddison, B.C.; Middleditch, D.J.; Patmore, J.R.M.; Gough, K.C. The detection of aquatic animal species using environmental DNA—A review of eDNA as a survey tool in ecology. *J. Appl. Ecol.* **2014**, *51*, 1450–1459. [CrossRef]
26. Kang, C.-K.; Kim, J.B.; Lee, K.-S.; Kim, J.B.; Lee, P.-Y.; Hong, J.-S. Trophic importance of benthic microalgae to macrozoobenthos in coastal bay systems in Korea: Dual stable C and N isotope analyses. *Mar. Ecol. Prog. Ser.* **2003**, *259*, 79–92. [CrossRef]
27. Lee, M.; Park, B.S.; Baek, S.H. Tidal Influences on Biotic and Abiotic Factors in the Seomjin River Estuary and Gwangyang Bay, Korea. *Estuaries Coasts* **2018**, *41*, 1977–1993. [CrossRef]
28. Korean Statistical Information Service (KOSIS). Available online: http://kosis.kr (accessed on 10 July 2018).
29. You, Y.-S.; Cho, H.-S.; Choi, Y.-C. A study on the pollution of polycyclic aromatic hydrocarbons (PAHs) in the surface sediments around Gwangyang Bay. *J. Korean Soc. Mar. Environ. Saf.* **2007**, *13*, 9–20.
30. Chihara, M.; Murano, M. *An Illustrated Guide to Marine Plankton in Japan*; Tokai University Press: Tokyo, Japan, 1997; p. 1574.
31. Albaina, A.; Aguirre, M.; Abad, D.; Santos, M.; Estonba, A. 18S rRNA V9 metabarcoding for diet characterization: A critical evaluation with two sympatric zooplanktivorous fish species. *Ecol. Evol.* **2016**, *6*, 1809–1824. [CrossRef]
32. Abad, D.; Albaina, A.; Aguirre, M.; Laza-Martínez, A.; Uriarte, I.; Iriarte, A.; Villate, F.; Estonba, A. Is metabarcoding suitable for estuarine plankton monitoring? A comparative study with microscopy. *Mar. Biol.* **2016**, *163*, 149. [CrossRef]
33. Guo, L.; Sui, Z.; Liu, Y. Quantitative analysis of dinoflagellates and diatoms community via Miseq sequencing of actin gene and v9 region of 18S rDNA. *Sci. Rep.* **2016**, *6*, 34709. [CrossRef]
34. Amaral-Zettler, L.A.; McCliment, E.A.; Ducklow, H.W.; Huse, S.M. A Method for Studying Protistan Diversity Using Massively Parallel Sequencing of V9 Hypervariable Regions of Small-Subunit Ribosomal RNA Genes. *PLoS ONE* **2009**, *4*, e6372. [CrossRef]
35. Edgar, R.C. Search and clustering orders of magnitude faster than BLAST. *Bioinformatics* **2010**, *26*, 2460–2461. [CrossRef]
36. Altschul, S.F.; Gish, W.; Miller, W.; Myers, E.W.; Lipman, D.J. Basic local alignment search tool. *J. Mol. Biol.* **1990**, *215*, 403–410. [CrossRef]
37. Caporaso, J.G.; Kuczynski, J.; Stombaugh, J.; Bittinger, K.; Bushman, F.D.; Costello, E.K.; Fierer, N.; Peña, A.G.; Goodrich, J.K.; Gordon, J.I.; et al. QIIME allows analysis of high-throughput community sequencing data. *Nat. Methods* **2010**, *7*, 335. [CrossRef]
38. Kohonen, T. *Self-Organizing Maps*; Springer: New York, NY, USA, 1997; p. 426.
39. Giraudel, J.L.; Lek, S. A comparison of self-organizing map algorithm and some conventional statistical methods for ecological community ordination. *Ecol. Model.* **2001**, *146*, 329–339. [CrossRef]
40. Chon, T.-S. Self-Organizing Maps applied to ecological sciences. *Ecol. Inform.* **2011**, *6*, 50–61. [CrossRef]
41. Kim, D.-K.; Jo, H.; Han, I.; Kwak, I.-S. Explicit Characterization of Spatial Heterogeneity Based on Water Quality, Sediment Contamination, and Ichthyofauna in a Riverine-to-Coastal Zone. *Int. J. Environ. Res. Public Health* **2019**, *16*, 409. [CrossRef]
42. Kim, D.-K.; Javed, A.; Yang, C.; Arhonditsis, G.B. Development of a mechanistic eutrophication model for wetland management: Sensitivity analysis of the interplay among phytoplankton, macrophtyes, and sediment nutrient release. *Ecol. Inform.* **2018**, *48*, 198–214. [CrossRef]
43. Chon, T.-S.; Park, Y.-S.; Cha, E.Y. Patterning of community changes in bentic macroinvertebrates collected from urbanized streams for the short term prediction by temporal artificial neuronal networks. In *Artificial Neuronal Networks: Application to Ecology and Evolution*; Lek, S., Guegan, J.F., Eds.; Springer: Berlin, Germany, 2000; pp. 99–114.

44. Park, Y.-S.; Kwon, Y.-S.; Hwang, S.-J.; Park, S. Characterizing effects of landscape and morphometric factors on water quality of reservoirs using a self-organizing map. *Environ. Model. Softw.* **2014**, *55*, 214–221. [CrossRef]
45. Kim, D.-K.; Jeong, K.-S.; Chang, K.-H.; La, G.-H.; Joo, G.-J.; Kim, H.-W. Patterning zooplankton communities in accordance with annual climatic conditions in a regulated river system (the Nakdong River, South Korea). *Int. Rev. Hydrobiol.* **2012**, *97*, 55–72. [CrossRef]
46. Hadjisolomou, E.; Stefanidis, K.; Papatheodorou, G.; Papastergiadou, E. Assessment of the Eutrophication-Related Environmental Parameters in Two Mediterranean Lakes by Integrating Statistical Techniques and Self-Organizing Maps. *Int. J. Environ. Res. Public Health* **2018**, *15*, 547. [CrossRef]
47. Várbíró, G.; Ács, É.; Borics, G.; Érces, K.; Fehér, G.; Grigorszky, I.; Japport, T.; Kocsis, G.; Krasznai, E.; Nagy, K.; et al. Use of self-organizing maps (SOM) for characterization of riverine phytoplankton associations in Hungary. *Arch. Hydrobiol.* **2007**, *161*, 388–394. [CrossRef]
48. Tison, J.; Giraudel, J.L.; Coste, M.; Park, Y.S.; Delmas, F. Use of unsupervised neural networks for ecoregional zoning of hydrosystems through diatom communities: Case study of Adour-Garonne watershed (France). *Arch. Hydrobiol.* **2004**, *159*, 409–422. [CrossRef]
49. Cuss, C.W.; Gueguen, C. Analysis of dissolved organic matter fluorescence using self-organizing maps: Mini-review and tutorial. *Anal. Methods* **2016**, *8*, 716–725. [CrossRef]
50. Penczak, T.; Głowacki, Ł.; Kruk, A.; Galicka, W. Implementation of a self-organizing map for investigation of impoundment impact on fish assemblages in a large, lowland river: Long-term study. *Ecol. Model.* **2012**, *227*, 64–71. [CrossRef]
51. Brosse, S.; Giraudel, J.L.; Lek, S. Utilisation of non-supervised neural networks and principal component analysis to study fish assemblages. *Ecol. Model.* **2001**, *146*, 159–166. [CrossRef]
52. Ha, J.-Y.; Hanazato, T.; Chang, K.-H.; Jeong, K.-S.; Kim, D.-K. Assessment of the lake biomanipulation by introducing both piscivorous rainbow trout and herbivorous daphnids using self-organizing map analysis: A case study in Lake Shirakaba, Japan. *Ecol. Inform.* **2015**, *29*, 182–191. [CrossRef]
53. Jeong, K.-S.; Kim, D.-K.; Pattnaik, A.; Bhatta, K.; Bhandari, B.; Joo, G.-J. Patterning limnological characteristics of the Chilika lagoon (India) using a self-organizing map. *Limnology* **2008**, *9*, 231–242. [CrossRef]
54. Vesanto, J.; Alhoniemi, E. Clustering of the Self-Organizing Map. *IEEE Trans. Neural Netw.* **2000**, *11*, 586–600. [CrossRef]
55. Shannon, C.E.; Weaver, W. *The Mathematical Theory of Communication*; The University of Illinois Press: Urbana, IL, USA, 1964; p. 125.
56. Vasselon, V.; Rimet, F.; Tapolczai, K.; Bouchez, A. Assessing ecological status with diatoms DNA metabarcoding: Scaling-up on a WFD monitoring network (Mayotte island, France). *Ecol. Indic.* **2017**, *82*, 1–12. [CrossRef]
57. Sun, C.; Zhao, Y.; Li, H.; Dong, Y.; MacIsaac, H.J.; Zhan, A. Unreliable quantitation of species abundance based on high-throughput sequencing data of zooplankton communities. *Aquat. Biol.* **2015**, *24*, 9–15. [CrossRef]
58. Kelly, R.P.; Port, J.A.; Yamahara, K.M.; Martone, R.G.; Lowell, N.; Thomsen, P.F.; Mach, M.E.; Bennett, M.; Prahler, E.; Caldwell, M.R.; et al. Harnessing DNA to improve environmental management. *Science* **2014**, *344*, 1455–1456. [CrossRef]
59. Goldberg, C.S.; Turner, C.R.; Deiner, K.; Klymus, K.E.; Thomsen, P.F.; Murphy, M.A.; Spear, S.F.; McKee, A.; Oyler-McCance, S.J.; Cornman, R.S.; et al. Critical considerations for the application of environmental DNA methods to detect aquatic species. *Methods Ecol. Evol.* **2016**, *7*, 1299–1307. [CrossRef]
60. Kembel, S.W.; Wu, M.; Eisen, J.A.; Green, J.L. Incorporating 16S Gene Copy Number Information Improves Estimates of Microbial Diversity and Abundance. *PLoS Comp. Biol.* **2012**, *8*, e1002743. [CrossRef]
61. Kelly, R.P.; Closek, C.J.; O'Donnell, J.L.; Kralj, J.E.; Shelton, A.O.; Samhouri, J.F. Genetic and Manual Survey Methods Yield Different and Complementary Views of an Ecosystem. *Front. Mar. Sci.* **2017**, *3*, 283. [CrossRef]
62. Stoeck, T.; Breiner, H.-W.; Filker, S.; Ostermaier, V.; Kammerlander, B.; Sonntag, B. A morphogenetic survey on ciliate plankton from a mountain lake pinpoints the necessity of lineage-specific barcode markers in microbial ecology. *Environ. Microbiol.* **2014**, *16*, 430–444. [CrossRef]
63. Gonzalez, J.M.; Portillo, M.C.; Belda-Ferre, P.; Mira, A. Amplification by PCR Artificially Reduces the Proportion of the Rare Biosphere in Microbial Communities. *PLoS ONE* **2012**, *7*, e29973. [CrossRef]

64. Elbrecht, V.; Leese, F. Can DNA-Based Ecosystem Assessments Quantify Species Abundance? Testing Primer Bias and Biomass—Sequence Relationships with an Innovative Metabarcoding Protocol. *PLoS ONE* **2015**, *10*, e0130324. [CrossRef]
65. Piñol, J.; San Andrés, V.; Clare, E.L.; Mir, G.; Symondson, W.O.C. A pragmatic approach to the analysis of diets of generalist predators: The use of next-generation sequencing with no blocking probes. *Mol. Ecol. Resour.* **2014**, *14*, 18–26. [CrossRef]
66. Chen, G.; Hare, M.P. Cryptic ecological diversification of a planktonic estuarine copepod. *Acartia Tonsa. Mol. Ecol.* **2008**, *17*, 1451–1468. [CrossRef]
67. Lindeque, P.K.; Parry, H.E.; Harmer, R.A.; Somerfield, P.J.; Atkinson, A. Next Generation Sequencing Reveals the Hidden Diversity of Zooplankton Assemblages. *PLoS ONE* **2013**, *8*, e81327. [CrossRef]
68. Hirai, J.; Kuriyama, M.; Ichikawa, T.; Hidaka, K.; Tsuda, A. A metagenetic approach for revealing community structure of marine planktonic copepods. *Mol. Ecol. Resour.* **2015**, *15*, 68–80. [CrossRef]
69. Massana, R.; Gobet, A.; Audic, S.; Bass, D.; Bittner, L.; Boutte, C.; Chambouvet, A.; Christen, R.; Claverie, J.-M.; Decelle, J.; et al. Marine protist diversity in European coastal waters and sediments as revealed by high-throughput sequencing. *Environ. Microbiol.* **2015**, *17*, 4035–4049. [CrossRef]
70. Coissac, E.; Riaz, T.; Puillandre, N. Bioinformatic challenges for DNA metabarcoding of plants and animals. *Mol. Ecol.* **2012**, *21*, 1834–1847. [CrossRef]
71. Eichmiller, J.J.; Best, S.E.; Sorensen, P.W. Effects of Temperature and Trophic State on Degradation of Environmental DNA in Lake Water. *Environ. Sci. Technol.* **2016**, *50*, 1859–1867. [CrossRef]

Article

Assessing Spatial Distribution of Benthic Macroinvertebrate Communities Associated with Surrounding Land Cover and Water Quality

Dong-Kyun Kim [1,2], Hyunbin Jo [1], Kiyun Park [1] and Ihn-Sil Kwak [1,3,*]

[1] Fisheries Science Institute, Chonnam National University, Yeosu 59626, Korea; dkkim1004@gmail.com (D.-K.K.); prozeva@hanmail.net (H.J.); ecoblue@hotmail.com (K.P.)
[2] K-water Research Institute, 1689beon-gil 125, Yuseongdaero, Daejeon 34045, Korea
[3] Faculty of Marine Technology, Chonnam National University, Yeosu 59626, Korea
[*] Correspondence: iskwak@chonnam.ac.kr; Tel.: +82-61-659-7148

Received: 29 September 2019; Accepted: 20 November 2019; Published: 28 November 2019

Abstract: The study aims to assess the spatial distribution of benthic macroinvertebrate communities in response to the surrounding environmental factors related to land use and water quality. A total of 124 sites were surveyed at the Seomjin River basin in May and September 2017, respectively. We evaluated the abundance and composition of benthic macroinvertebrate communities based on nine subwatersheds. Subsequently, we compared the benthic information with the corresponding land use and water quality. To comprehensively explore the spatiotemporal distinction of benthic macroinvertebrate communities associated with those ambient conditions, we applied canonical correspondence analysis (CCA). The CCA results explicitly accounted for 61% of the explanatory variability; the first axis (45.5%) was related to land-use factors, and the second axis (15.5%) was related to water quality. As a result, the groups of benthic communities were distinctly characterized in relation to these two factors. It was found that land-use information is primarily an efficient proxy of ambient water quality conditions to determine benthic macroinvertebrates, such as *Asellus* spp., *Gammarus* spp., and *Simulium* spp. in a stream ecosystem. We also found that specific benthic families or genera within the same groups (Coleoptera, Diptera, Ephemeroptera, and Trichoptera) are also differentiated from ambient water quality changes as a secondary component. In particular, the latter pattern appeared to be closely associated with the impact of summer rainfall on the benthic community changes. Our study sheds light upon projecting benthic community structure in response to changes of land use and water quality. Finally, we conclude that easily accessible information, such as land-use data, aids in effectively characterizing the distribution of benthic macroinvertebrates, and thus enables us to rapidly assess stream health and integrity.

Keywords: benthic macroinvertebrates; canonical correspondence analysis; land use; spatial distribution; water quality

---

1. Introduction

Benthic species are one of the most diverse and abundant biota in fluvial ecosystems such as rivers and streams [1]. Recognizing a large portion of their importance in fluvial ecosystems, the ecological responses of those benthic species to ambient physicochemical conditions have been explored and described for the sake of biological assessments based on species sensitivity [2–4]. For many years, it has been thought that benthic macroinvertebrate communities generally play a pivotal role in facilitating energy flows and nutrient cycling within ecosystems [5,6]. McLenaghan et al. [7] reported that the functional diversity of benthic macroinvertebrates regulates nutrient and algal dynamics in riverine ecosystems. Besides, the high sensitivity of species in their composition and assemblage to changes of ambient habitat conditions allows benthic macroinvertebrates to be used for assessing the stream health and integrity of fluvial ecosystems [8,9]. From an ecological perspective related to niche partitioning, monitoring the distribution of aquatic macroinvertebrates has been linked to ambient physicochemical constraints (e.g., ecosystem morphology and trophic status) [10,11]. Hence, the role of benthic invertebrates has been gradually emphasized as bioindicators [12].

Stream health based on benthic communities can be spatially heterogeneous according to ambient environmental factors, such as neighboring land use/cover, various pollutants, hydrological factors, and local climates. Particularly, land use/cover is a critical factor to drive the transport of sediments and nutrients related to stream water quality [13,14]. Given the recently advanced satellite technology, easily obtainable/accessible data to land-use information are highly cost-efficient relative to field-based water quality measurement [14,15]. Since the fate and transport of nutrients (e.g., nitrogen and phosphorus) are also closely associated with land use in watersheds, benthic communities can be correlated with surrounding land-use patterns [16,17]. Therefore, we hypothesize that land-use information from online websites can enable the rapid assessment of benthic communities in the context of ecosystem health.

Despite the assessment efficiency of land-use information, its slow changes can be still limited to explicitly account for the temporal dynamics of the target biota of our interest within a short term. In stream ecosystems, benthic macroinvertebrates vary in their composition as well as abundance within the same survey area over time. Particularly in East Asian countries including Korea, eastern China, and Japan, monsoon events along with multiple typhoons recur in summer [18,19]. Despite the short time span between surveys, this local climatological feature can change water quality quickly. Therefore, we infer that there are huge potentials in benthic macroinvertebrate community change within a short term, assuming that ambient water quality can be a supplementary indicator to characterize the benthic macroinvertebrate community in a temporal scale.

Our study aims to project the distributions of benthic macroinvertebrate communities associated with the surrounding land use/cover and water quality in Seomjin River, South Korea. We also analyze and evaluate the sensitivities of benthic communities in different taxonomical levels (e.g., order, genus, and species). Furthermore, to assess the status of stream health and integrity in the Seomjin River, we evaluated the values of biotic indices from the collected benthic macroinvertebrate data. Finally, we anticipate finding out more useful data information to effectively characterize the distribution of benthic macroinvertebrates in the context of the rapid assessment of stream health.

2. Materials and Methods

2.1. Site Description

Seomjin River is one of the four major watersheds in South Korea and is located in the southwestern part of the Korean peninsula (34°55′–35°45′ N, 126°57′–127°55′ E) (Figure 1). The catchment area is approximately 4900 km^2, and the length of the river is 223 km. The primary land use is a forest area (48%), and the second dominant land use is an agricultural area (33.6%) (Figure 1). The urban lands comprise approximately 4% of the total catchment area. The agricultural lands are mainly adjacent to stream channels. The annual precipitation has been 1384 ± 317 mm (mean ± SD) in the Seomjin River catchment for the last decade. The unique climate conditions, such as monsoon and typhoons, bring about >50% of the precipitation concentrated in summer between June and September (Figure 2). For this reason, several multi-purpose dams are operational to effectively manage water resources for irrigation, potable use, and flow regulation. The Korean Ministry of the Environment manages the Seomjin River watershed, dividing it into nine subwatersheds in compliance with their geophysical characteristics, specifically the main channel, tributaries, upstream, downstream, and dam location. Particularly for the two groups of subwatersheds; (i) Seomjin Dam in the upper site (SDU) and the Seomjin Dam in the lower site (SDL), and (ii) Juam Dam (JD) and Bo-Seong (BS), the main characteristics of these subwatersheds are separated by dam location. The former subwatersheds form part of the headwater of the main channel, while the latter are of the main tributary (Figure 1).

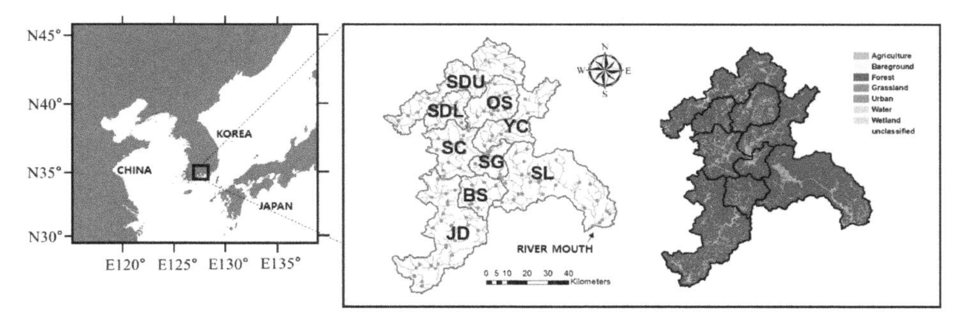

Figure 1. Map of the study area and land-use information in the Seomjin River catchment. The red circles indicate a total of 124 study sites. The entire watershed was divided into nine subwatersheds.

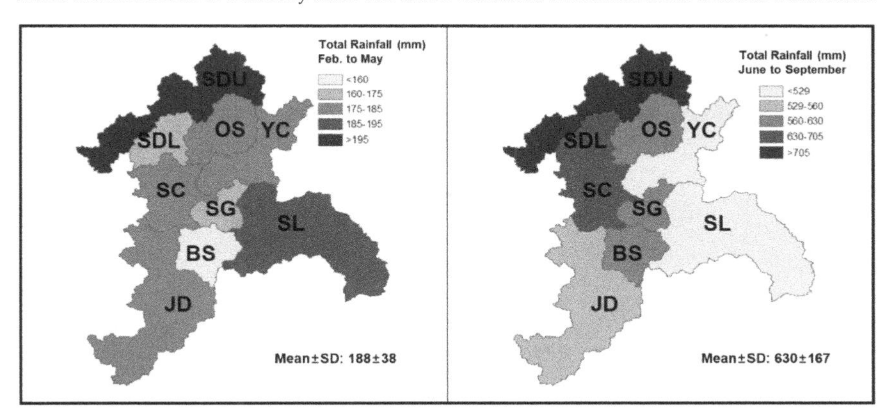

Figure 2. Total precipitation of the Seomjin River basin in 2017, based on nine major subwatersheds.

2.2. Data Collection

The present study was based on 124 sites across the Seomjin River watershed in 2017. The study sites are part of the national water quality monitoring networks run by the National Institute of Environmental Research (NIER) which is operated by the Korean Ministry of Environment (KMOE). All 124 sites are located in the same catchment and are connected to one another by way of the Seomjin River. Due to the accessibility of sediment sampling, most of the study sites are placed on low-order streams (i.e., shallow depth). For data consistency, we conducted the surveys at the same sites twice a year in May and September. The total number of study sites consists of 16 from SDU (catchment area: 763 km^2), five from SDL (237 km^2), 12 from Oh-Soo, OS (370 km^2), 12 from Soon-Chang, SC (431 km^2), 11 from Yo-Cheon, YC (486 km^2), seven from Seomjin-Gokseong, SG (183 km^2), 27 from the lower Seomjin River, SL (1128 km^2), 24 from JD (1029 km^2), and 10 from BS (283 km^2) (Figure 1).

From the study sites, we investigated geophysicochemical features, such as land-use information and water quality parameters. The land-use data were based on the year 2016, and were obtained from the National Spatial Data Infrastructure Portal (http://data.nsdi.go.kr). We specifically extracted the land-use data around the study sites by an arbitrary 1-km circle buffer using ArcGIS software (ESRI, Redlands, CA, USA). We collected water samples on each site (one sample per site). The water quality parameters included biochemical oxygen demand (BOD), total nitrogen (TN), nitrate (NO_3-N), ammonia (NH_3-N), total phosphorus (TP), phosphate (PO_4-P), and chlorophyll *a* concentrations (Chl-*a*).

The water quality parameters, including BOD, TN, NO_3-N, NH_3-N, TP PO_4-P, and Chl-*a* concentrations were analyzed in the laboratory using water samples on sites in compliance with the methods proposed by Wetzel and Likens [20].

For biological data, we sampled three benthic sediments at each site, taking the spatial heterogeneity within the site into account. A Surber sampler (30 cm × 30 cm, 500 μm mesh; APHA et al., 1992) was used to collect benthic macroinvertebrates, at a depth of approximately 10 cm in May and September. Then, we preserved the obtained benthic macroinvertebrates in 7% formalin. In the laboratory, we sorted the invertebrate specimens, identified them up to genus or species level, and counted the number of specimens using a dissecting anatomy microscope. The identification was based on several pieces of literature including Quigley [21], Pennak [22], Brighnam et al. [23], Yun [24], and Merritt and Cummins [25].

2.3. Use of Biotic Indices

To assess the status of stream health and integrity in the Seomjin River, we calculated the values of biotic indices from the collected benthic macroinvertebrate data. Bearing in mind that there is no clear-cut distinction of ecosystem assessment in accordance with biotic indices, we considered both globally popular and regionally specific indices. In this respect, five biotic indices were selected to evaluate the abundance, diversity, dominance, evenness, and richness: McNaughton's dominance index (DI, [26]), Shannon–Weaver index (H', [27]), Richness index (RI, [28]), Evenness index (EI, [29]), and Benthic Macroinvertebrates index (BMI; [30]). In particular, BMI is a modified version (i.e., conceptually the same) of the saprobic index of Zelinka and Marvan [31], which ranges from 0 to 100. The BMI has been used for the bio-assessment of benthic macroinvertebrates in Korea [32]. The governing equation is expressed as:

$$BMI = \left(4 - \frac{\sum_{i=1}^{n} s_i h_i g_i}{\sum_{i=1}^{n} h_i g_i}\right) \times 25 \tag{1}$$

where s_i denotes the saprobic value of the species i, h_i denotes the relative abundance ranking of the species i, and g_i denotes the weight value of the species i of the total number of species n. There is a subtle difference between the saprobic index and BMI. While the saprobic index takes the absolute biomass of the species for h_i, BMI uses the relative ranking of species abundance. Kong et al. [30] reported that BMI was a more capable means of assessing stream health and integrity when utilizing the information of relative abundance.

2.4. Multivariate Analysis for Data Ordination

We used canonical correspondence analysis (CCA) in order to relate the benthic macroinvertebrate communities to the surrounding environmental variables. The environmental variables included the land-use percent coverage of cropland, urban land, forest, and wetland, in addition to the ambient water quality parameters TN, TP, NH3, Chl-a, and BOD. The benthic macroinvertebrate data included 22 genus groups, except for the family group of Chironomidae. Prior to the CCA application, we calculated the length of the gradient based on detrended correspondence analysis in order to examine the adequacy of CCA application [33]. As the length was 4.16 (greater than 4), CCA was used for data ordination [34,35]. All the variables were modified by log transformation to stabilize the data close to normal distribution [36]. For statistical analysis, we used SPSS ver. 18 software (IBM, New York, NY, USA) and the open-source software PAST3 (SOFTPEDIA, Romania).

3. Results and Discussion

3.1. Comparison of Water Quality and Biotic Indices

Water quality parameters and biotic indices were compared in nine subwatersheds between May and September. In the temporal scale, the nitrogen level was slightly higher in May than in September (Figure 3). It appeared that a significant amount of nutrients entered the stream by summer rainfall (see Figures 2 and 3). We speculate that incoming nutrients are more dissolved forms of nitrogen and phosphorus than particulate, because we observed frequent increases of NH_3–N, NO_3–N, and PO_4–P (Figure 3). The previous studies have reported that the high flux of NO_3–N could be driven by specific agricultural activities, such as manure applications and conservational tillage [37,38]. Nonetheless, TN concentrations exhibited a gradual decrease from upstream to downstream (Figure 3), and specifically exhibited approximately from 2 mg L^{-1} at sites SDU, SDL, and OS, to 1 mg L^{-1} at sites SG and SL. A study by Ahn [39] reported that unexpectedly high nutrient concentrations in headstreams are often observed because of the relatively undeveloped wastewater treatment and septic tanks in the rural area of South Korea, thereby inducing an increase of BOD. In contrast to TN and NO_3–N, NH_3–N concentrations were very high at the SC site, regardless of time (Figure 3). Given the largest fraction (49%) of cropland among nine subwatersheds, the high ammonia concentrations could be associated with agricultural activities. In China, which is geographically close to Korea, it was reported that agricultural nitrogen accounted for more than 40% of the variability of TN, and subsequently it drives ammonia increases [40,41].

The phosphorus pattern was spatially similar to the nitrogen pattern in May. The level of TP concentrations gradually decreased from upstream to downstream, except at the headstream SDU site in May (Figure 3). However, it was comparatively notable that phosphorus concentrations rebounded in the lower part of the river, such as at the BS and SL sites. This longitudinal trend could be consistent with the fact that the lower part of the Seomjin River was dominated by agricultural lands (38.8% at the IS, SC, YC, BS, and SL sites) and urban areas (6.6% at the YC and BS sites) relative to the upper part (27% at the SDU, SDL, SG, and JD sites, and 3.3% at the SDU, SDL, and JD sites) (see Figure 1). In relation to agricultural management practices (e.g., manure applications) and urbanization, a large amount of dissolved phosphorus could be generated [42,43].

Chl-*a* concentrations were also slightly higher at the upper part than at the lower part of the river. However, based on the current level of Chl-*a* concentrations, the Seomjin River changed from being oligotrophic to mesotrophic (Chl-*a* < 10 μg L^{-1}). The BOD values mostly ranged from 1 to 3 mg L^{-1}, and appear to differ from the spatial pattern of nutrient concentrations (Figure 3). Additionally, we also compared five biotic indices, and most of the values of biotic indices were not statistically significant across all the study sites (Figure 3). Thus, it was difficult to distinguish the spatial pattern of benthic communities based merely on the biotic indices.

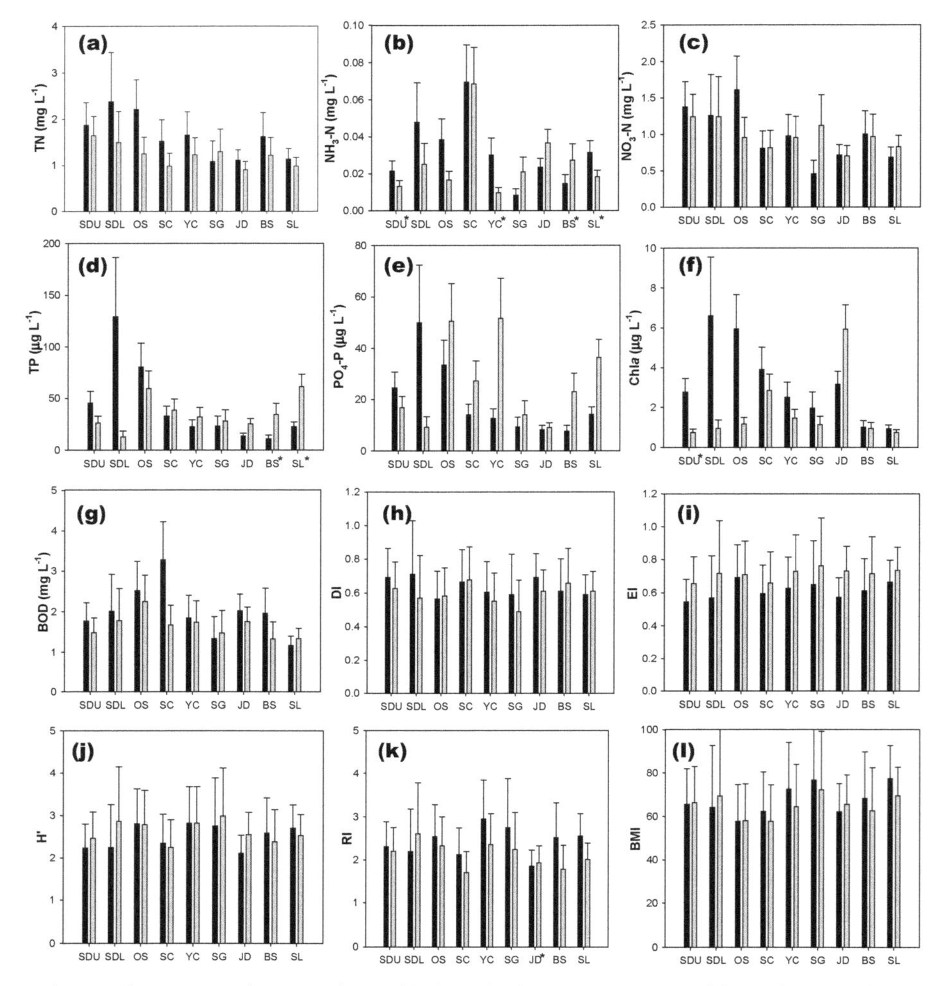

Figure 3. Comparison of water quality and biological indices between May and September. Error bars indicate standard errors. Asterisks represent statistical significances at $P < 0.05$. (**a**) TN; (**b**) NH_3-N; (**c**) NO_3-N; (**d**) TP; (**e**) PO_4-P; (**f**) Chl-*a*; (**g**) BOD; (**h**) DI; (**i**) EI; (**j**) H′; (**k**) RI; (**l**) BMI.

3.2. Spatial Distribution of Benthic Macroinvertebrate Communities Before and After Summer Rainfall

The distribution of benthic macroinvertebrate communities was spatially distinct (Figure 4). At all nine of the subwatersheds, the abundance of benthic macroinvertebrates decreased after the summer rainfall in September. A decreasing level of benthic communities differed from the location of subwatersheds. It was a particularly remarkable pattern that while the abundance of *Gammarus* spp. (Amphipoda) slightly increased, the abundance of *Asellus* spp. (Isopoda) dramatically decreased after the heavy rainfall (Figure 4). Moreover, the relative abundance decreased from 41% to 3.3% (Table 1). However, in this respect, *Gammarus* spp. also dramatically increased in terms of relative abundance (26.6% in May to 69.6% in September, Table 1). This pattern was clearer at the SDU site, which was a forest-dominated area. We could not find evidence to support their inverse abundance pattern associated with rainfall, especially in a forest area. To understand their relationships and ecological interactions, a long-term monitoring is highly required to depict the inter-annual variation in specific land-use coverage.

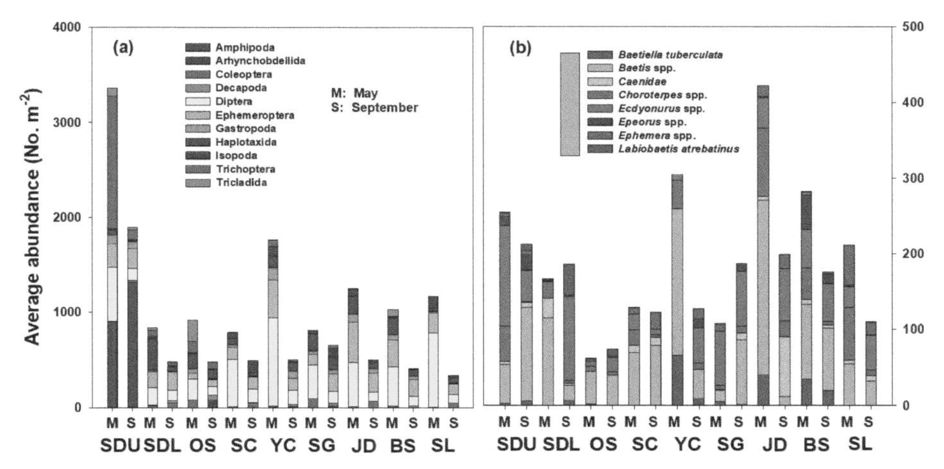

Figure 4. Average abundance of benthic macroinvertebrate communities based on (**a**) order and (**b**) subcategory (family, genus, and species) in Seomjin River.

In relation to unfavorable benthic macroinvertebrates in polluted environments, it was found that Diptera (Chironomidae and *Simulium* spp.) prevailed across the Seomjin River watershed (Figure 3). However, it was also observed that their abundance was significantly low in September after summer rainfall. Chironomidae abundance consistently decreased across all of the study sites except for SDL. A slight increase (21.9% to 24.1%) of the abundance might be longitudinal flush effects, since the SDL site was located downstream of the Seomjin Dam. *Simulium* spp. also decreased after summer rainfall, but their low abundance appeals to further surveys over a long term.

On the other hand, the key benthic macroinvertebrates such as Coleoptera and Ephemeroptera were scrutinized. Interestingly, the abundance patterns of *Elmidae* spp. and *Eubrianax* spp. were opposite between May and September. The former was commonly higher in May, while the latter was higher in September (Table 1). Ephemeroptera were slightly more abundant in September. Particularly, *Ecdyonurus* spp. abundance was distinctly high in September relative to the other Ephemeroptera (Table 1). The most dominant Ephemeroptera, *Baetis* spp., showed irregular spatial pattern in their abundance. Interestingly, the longitudinal pattern of abundance looked like a zigzag, which implies that *Baetis* spp. (e.g., larva) could be influenced serially from upstream to downstream by hydrological factors. Another key group Trichoptera showed higher abundance, particularly at the SDL and OS sites (Table 1). *Hydropsyche* spp. appeared to be sensitive to summer rainfall, while *Cheumatopsyche* spp. seem to be more tolerant.

Table 1. The relative abundances of benthic macroinvertebrates in Seomjin River. SDU: Seomjin Dam in the upper site, SDL: Seomjin Dam in the lower site, OS: Oh-Soo, SC: Soon-Chang, YC: Yo-Cheon, SG: Seomjin-Gokseong, JD: Juam Dam, BS: Bo-Seong, SL: Seomjin River.

Order	Family or Genus	Month	SDU	SDL	OS	SC	YC	SG	JD	BS	SL
Amphipoda	*Gammarus* spp.	May	26.6%	0.0%	0.0%	0.0%	0.1%	0.0%	0.0%	0.0%	0.6%
		Sep	69.6%	0.0%	0.0%	0.0%	0.1%	0.0%	0.0%	0.0%	0.0%
Arhynchobdellida	*Erpobdella* spp.	May	0.1%	0.0%	0.3%	0.2%	0.3%	0.2%	0.1%	0.1%	0.3%
		Sep	0.0%	0.2%	12.0%	2.1%	2.7%	2.0%	1.0%	0.2%	0.8%
Coleoptera	*Elmidae* spp.	May	0.2%	1.9%	8.3%	0.7%	0.6%	10.6%	0.5%	0.5%	0.3%
		Sep	0.0%	9.0%	2.4%	0.3%	2.9%	0.1%	1.9%	0.7%	1.7%
	Eubrianax spp.	May	0.0%	0.0%	0.0%	0.1%	0.1%	0.1%	0.0%	0.1%	0.5%
		Sep	0.9%	0.6%	1.2%	6.6%	0.5%	4.2%	9.6%	11.8%	1.1%
Decapoda	*Caridina* spp.	May	0.0%	0.5%	0.1%	0.3%	0.0%	0.1%	0.0%	0.1%	0.0%
		Sep	0.0%	4.5%	10.9%	1.1%	0.3%	0.2%	0.0%	0.7%	0.3%
Diptera	Chironomidae	May	9.6%	21.9%	23.5%	61.7%	51.6%	44.3%	66.3%	36.1%	37.8%
		Sep	5.9%	24.1%	19.5%	30.3%	30.3%	19.1%	28.1%	19.5%	24.4%
	Simulium spp.	May	7.4%	0.5%	0.4%	0.7%	0.5%	0.3%	0.1%	0.7%	1.9%
		Sep	0.7%	0.0%	0.3%	0.0%	0.0%	0.1%	0.2%	0.0%	1.1%
Ephemeroptera	*Baetiella* spp.	May	0.1%	0.0%	0.2%	0.0%	3.8%	0.7%	0.0%	3.3%	3.4%
		Sep	0.3%	1.5%	0.1%	0.0%	2.0%	0.4%	0.0%	0.0%	5.1%
	Baetis spp.	May	1.5%	13.7%	4.6%	8.8%	10.9%	1.8%	4.7%	18.4%	9.5%
		Sep	6.5%	4.0%	8.2%	16.1%	7.6%	12.8%	9.7%	2.4%	20.2%
	Caenidae	May	0.1%	3.2%	0.0%	1.1%	0.0%	0.1%	0.4%	0.4%	0.7%
		Sep	0.4%	0.6%	0.0%	2.2%	0.0%	1.5%	1.8%	15.9%	0.9%
	Choroterpes spp.	May	1.4%	2.6%	0.7%	2.7%	2.1%	0.8%	6.0%	7.2%	4.0%
		Sep	0.0%	0.8%	0.9%	0.7%	1.6%	1.3%	2.4%	4.0%	1.1%
	Ecdyonurus spp.	May	4.0%	0.0%	0.7%	2.6%	2.9%	8.7%	2.3%	3.2%	5.0%
		Sep	2.2%	23.1%	3.9%	1.4%	9.3%	11.2%	13.5%	14.0%	12.4%
	Epeorus spp.	May	0.4%	0.1%	0.0%	0.0%	1.7%	0.1%	0.3%	0.1%	4.4%
		Sep	1.1%	0.0%	0.1%	0.1%	2.4%	1.5%	0.1%	0.0%	3.2%
	Ephemera spp.	May	0.1%	0.2%	0.4%	1.1%	1.4%	1.1%	4.5%	1.2%	0.4%
		Sep	0.3%	0.3%	0.3%	4.5%	2.7%	0.1%	4.9%	3.8%	0.6%
	Labiobaetis spp.	May	0.0%	0.3%	0.0%	0.1%	0.0%	0.0%	0.0%	0.0%	0.0%
		Sep	0.5%	8.7%	1.9%	0.0%	0.1%	0.0%	0.2%	0.0%	0.1%

Table 1. *Cont.*

Order	Family or Genus	Month	SDU	SDL	OS	SC	YC	SG	JD	BS	SL
Gastropoda	*Semisulcospira* spp.	May	1.5%	40.4%	16.6%	8.1%	7.0%	22.2%	3.0%	15.3%	16.3%
		Sep	0.6%	11.6%	19.4%	7.1%	18.1%	19.6%	16.2%	11.8%	15.6%
	Radix spp.	May	0.4%	0.1%	2.1%	0.7%	1.3%	1.7%	0.3%	0.3%	2.0%
		Sep	0.1%	0.0%	0.3%	0.0%	0.2%	1.0%	0.8%	6.6%	1.4%
Haplotaxida	*Limnodrilus* spp.	May	0.5%	1.7%	2.1%	7.7%	6.2%	3.1%	9.3%	5.4%	1.5%
		Sep	0.5%	2.9%	1.7%	25.6%	2.2%	15.2%	6.9%	6.0%	2.6%
Isopoda	*Asellus* sp.	May	41.0%	0.0%	2.1%	0.2%	0.5%	1.1%	1.2%	0.0%	0.0%
		Sep	3.3%	0.2%	0.6%	0.0%	0.1%	4.8%	0.0%	0.0%	0.1%
Trichoptera	*Cheumatopsyche* spp.	May	0.2%	8.4%	10.4%	0.1%	2.6%	0.0%	0.0%	0.1%	0.5%
		Sep	2.2%	5.0%	13.6%	0.0%	0.2%	0.0%	0.2%	0.0%	0.4%
	Hydropsyche spp.	May	2.6%	3.2%	24.5%	0.2%	0.6%	0.1%	0.0%	0.7%	7.4%
		Sep	1.2%	2.9%	1.7%	0.0%	2.4%	0.0%	0.0%	0.0%	0.1%
Tricladida	*Dugesia* spp.	May	2.3%	1.3%	2.9%	3.0%	5.9%	2.9%	0.9%	6.8%	3.5%
		Sep	3.7%	0.0%	0.9%	1.8%	14.3%	5.1%	2.5%	2.8%	6.6%

3.3. Association among Benthic Macroinvertebrates, Land-Use Coverage, and Ambient Water Quality

The CCA simplified the relationships among the variables of our interest (Figure 5). The CCA results explicitly accounted for 61% of the relational variability with two primary ordination axes; the first axis (45.5%) was related to land-use factors, and the second axis (15.5%) appeared to be related to water quality. The first ordination axis characterized the gradient of land-use coverage (Figure 5). The land-cover gradient was mainly separated by forest and agricultural/cropland areas. Urban land is topographically placed in the middle between forest and cropland. Wetland is close to cropland, which is reasonable because extensive agricultural areas have been converted from riverine wetlands by constructing levees in South Korea [44,45]. The second ordination axis depicted the gradient of water quality parameters. On the whole, the upper part of plot is closely associated with the parameters related to water quality deterioration, such as the higher concentration level of BOD, Chl-*a*, TP, and TN (Figure 5). It was notable that the ammonia (NH_3–N) concentration showed an inverse pattern against the other water quality parameters. This pattern appears to be associated with agriculture and pasture. In contrast, the NO_3–N concentration showed a weak relationship with agricultural activities including grassland/pasture, while the PO_4–P concentration had close association (Figures 5 and 6a). As previous mentioned, higher PO_4–P concentrations were apparently observed in the lower part of the river (Figure 3). Thus, the CCA ordination depicted the disparate responses of nitrogen and phosphorus dynamics in the lower part of the Seomjin River.

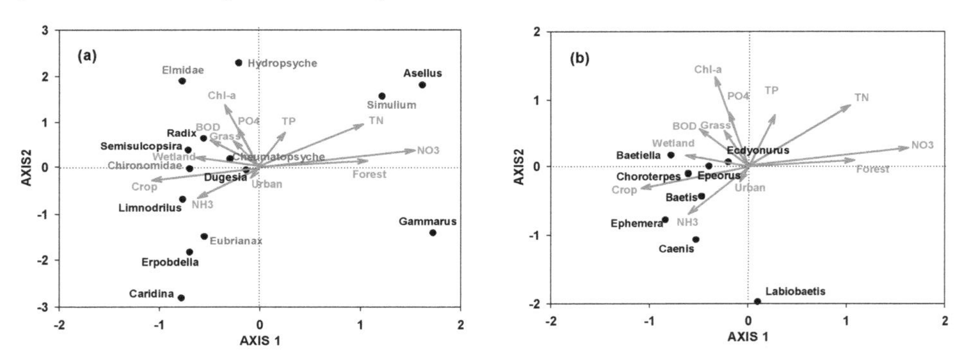

Figure 5. Results of data ordination based on canonical correspondence analysis (CCA). Comparison of land use and water quality with benthic macroinvertebrate communities; (**a**) 14 groups, and (**b**) eight Ephemeroptera groups.

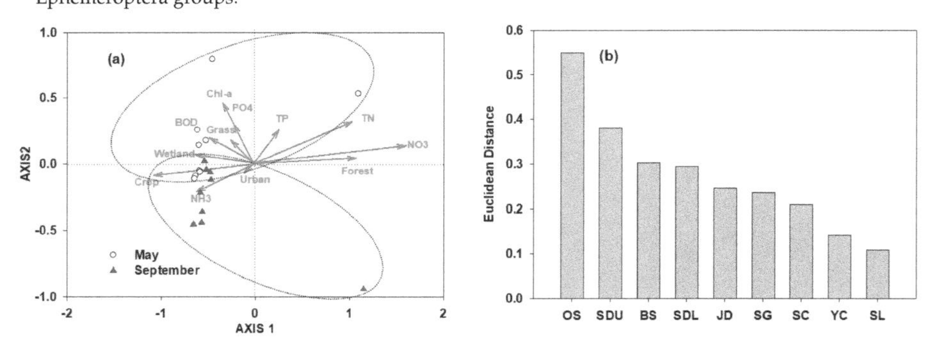

Figure 6. (**a**) CCA data ordination associated with data samples and ambient environmental factors. (**b**) Dissimilarity of data attributes between May and September.

In the ordination pattern of key benthic macroinvertebrates (e.g., EPTC taxa: E = Ephemeroptera, P = Plecoptera, T = Tricoptera, and C = Coleoptera), the major Coleoptera groups, *Elmidae* spp.

and *Eubrianax* spp., were distinctly differentiated according to water quality. These two genera, especially the riffle beetles *Elmidae*, are typically known as a large group of aquatic Coleoptera that are generally indicators of good water quality, because of their sensitivity to changes in the surrounding environmental conditions [46,47]. Nonetheless, the association of *Elmidae* with several water quality signals to eutrophic conditions (PO_4-P, Chl-*a*, and BOD) was unexpected (Figure 5a). This counterintuitive pattern may be a confounding effect stemming from the spatial migration of larvae. However, there have been some evidence that *Elmidae* can be distributed in a wide range of nutrient conditions [48]. We speculate that *Elmidae* can prevail to some extent, since the Seomjin River watershed is mainly dominated by forest (48% on average, see Figure 1). The Tripcoptera groups, *Hydropsyche* spp. and *Cheumatopsyche* spp., were separately characterized. It appears that *Hydropsyche* is more related to Chl-*a* than *Cheumatopsyche*. There have been several literatures on food preferences and niche partitioning among these species [49,50]. We understand the separate pattern of these species, in the sense that Hydropsyche larvae prefer higher-velocity microhabitats and mainly digest detritus and benthic diatoms [51]. Another key benthic macroinvertebrate Ephemeroptera generally tended to inhabit good water quality conditions (Figure 5b). However, they seemed to be apart from mountainous areas (i.e., forest areas). Within the same family Baetidae, *Baetiella* spp. was more associated with wetland habitat, while *Baetis* spp. was more associated with ammonia concentration. Particularly, it was found that *Baetis* spp. was less sensitive to ammonia toxicity than other mayflies [52]. We infer that the linkage between *Baetis* spp. and ammonia concentration is highly associated with their tolerance. In this context, *Ephemera* spp. and *Caenis* spp. seem to inhabit in a similar water quality condition (Figure 5b). The spatial ordination of *Labiobaetis* spp. was clearly distinct, which appears to be related to good water quality. The other Ephemeroptera, such as *Choropterpes* spp., *Epeorus* spp., and *Ecdyonurus* spp., were placed on a mixture of wetland, cropland, and urban areas. However, there was little evidence to advocate their association with surrounding environmental conditions.

The dominant Diptera groups were Chironomidae and *Simulium* spp. in the Seomjin River basin. These two groups were clearly separated in the data ordination induced by land-use coverage rather than water quality (Figure 5a). Chironomidae were closely associated with wetland and cropland. Plenty of literature papers have reported that Chironomidae are the most abundant insects in wetlands and play an important role in wetland food webs [53,54]. In addition, it is clear that agricultural land use has influenced water quality, as evidenced by high nutrient concentrations [55]. Thus, the accumulation of organic matters subsequently fosters the colonization of Chironomidae groups, particularly in lowland agricultural areas [55,56]. In contrast, the same group of Diptera, *Simulium* spp., was correlated with nitrogen rather than phosphorus in forest areas. The aforementioned *Hydropsyche* spp. is known to prefer boulder microhabitats, which are commonly found in upstream areas (i.e., forest area in Seomjin River). It was remarkable to show a weak relationship between *Hydropsyche* spp. and forest land cover. Given the interspecific competition between *Hydropsyche* spp. and *Simulium* spp., our result was plausible to make *Hydropsyche* spp. separate in order to avoid excessive predation and competition [57].

Of the remaining benthic macroinvertebrates, *Asellus* spp. and *Gammarus* spp. were strongly identified in forest areas (Figure 5a). These two genera were clearly separated by components related to water quality, which we will discuss in relation to temporal water quality changes in the following sections. The Gastropoda groups *Radix* spp. and *Semisulcopsira* spp. were similarly ordinated in the plot (Figure 5a). Although the former was more related to grassland (i.e., pasture) and the latter was more related to wetlands, it was difficult to characterize their habitat preferences. In this regard, further investigations are required. The Haplotaxida *Limnodrilus* spp. were closely associated with ammonia concentration in agricultural areas, and are known to be tolerant of unfavorable condition such as hypoxic and eutrophic states [58]. Thus, we suppose that *Limnodrilus* spp. is less sensitive to ammonia (NH_3–N) toxicity, similar to the aforementioned *Baetis* spp. and *Ephemera* spp.

3.4. Identification of Spatiotemporal Characteristics in the Data from Seomjin River

We portrayed the data ordination according to time and space (Figure 6a). It was remarkable to primarily characterize the data characteristics between May and September. Two data points were associated with forest areas. In comparison with Figure 5a, it was certain that these data points were correlated with *Asellus* spp. (Isopoda) and *Gammarus* spp. (Amphipoda), respectively. However, most data points characterized their clear separation across the vertical axis, which implies that benthic macroinvertebrate communities were affected by the temporal change of water quality. Through this pattern, the key macroinvertebrates Coleoptera *Elmidae* spp. and Tricoptera *Hydropsyche* spp. were relatively abundant in May, while Coleoptera *Eubrianax* spp. were relatively abundant in September (Figures 5a and 6a). Interestingly, the Ephemeroptera groups were closely associated with September compared to other key macroinvertebrates (Figures 5b and 6a).

We also question what major factors could drive the water quality changes between May and September. Given the unique climatological features of East Asia, such as summer monsoons and typhoons [18,19], we conjecture that a significant factor of water quality change would be the precipitation between two time periods (Figure 2). Since our CCA did not accommodate the precipitation data, we put more emphasis on the intensity of rainfall as a key factor for benthic macroinvertebrate communities, especially in East Asian countries. There have also been plenty of literatures related to the effects of flooding on benthic macroinvertebrates on a global scale [59,60].

Keeping this clear temporal pattern of data ordination in mind, we estimated the dissimilarity of data samples between May and September. Since the amount of rainfall was spatially distinct and the corresponding land-use coverage differs, there was a conspicuous deviation of data ordination between May and September (Figure 6b). The OS and SDU sites showed a larger disparity of data compared to the YC and SL sites. While the SDU site is in a headstream area, the SL site is near the river mouth (Figure 1). In this respect, their sensitivity to rainfall could be distinct. This pattern was clearer at the SDU site, which is a forest-dominated area and was consistent with the data ordination from CCA (Figure 5).

4. Conclusions

From this study, we demonstrated how spatially benthic macroinvertebrate communities were closely related to surrounding environmental constraints such as the surrounding land use and ambient water quality. The present study depicted that land-use coverage is a primary factor and water quality is a secondary factor to evaluate benthic macroinvertebrate communities. Our analysis also showed that the water quality change in the Seomjin River basin was mainly influenced by summer precipitation, thereby inducing a community shift of benthic macroinvertebrates in Korea. In addition, we estimated and compared quantitatively the influence of summer rainfall on a spatial scale, and then linked those deviations with the surrounding land-use coverage. The data ordination explicitly accounted for 61% of the explanatory variability in benthic macroinvertebrate communities. We stress that land-use information is primarily an efficient proxy of ambient conditions to determine benthic macroinvertebrates in a stream ecosystem. Finally, our study highlights that land-use information, which is easily obtainable, is very helpful for delineating the spatial distribution of benthic macroinvertebrate communities in stream ecosystems.

Author Contributions: Conceptualization: D.-K.K., H.J., K.P., and I.-S.K.; Methodology: D.-K.K. and H.J.; Formal Analysis: D.-K.K. and H.J.; Investigation: H.J. and K.P.; Resources: I.-S.K.; Writing—Original Draft Preparation: D.-K.K.; Writing—Review and Editing: D.-K.K. and I.-S.K.; Supervision: I.-S.K.; Project Administration: I.-S.K.; Funding Acquisition: I.-S.K.

Funding: This research was funded by the National Research Foundation (NRF) of Korea, grant number NRF-2018R1A6A1A03024314, and was also supported by the project 'Stream/River Ecosystem Survey and Health Assessment of Korea Ministry of Environment (KMOE).

Conflicts of Interest: The authors declare no conflict of interest.

References

1. Rosenberg, D.M.; Resh, V.H. *Freshwater Biomonitoring and Benthic Macroinvertebrates*; Springer: Berlin/Heidelberg, Germany, 1993; p. 488.
2. Baird, D.J.; Van den Brink, P.J. Using biological traits to predict species sensitivity to toxic substances. *Ecotoxicol. Environ. Saf.* **2007**, *67*, 296–301. [CrossRef] [PubMed]
3. Doledec, S.; Statzner, B. Invertebrate traits for the biomonitoring of large European rivers: An assessment of specific types of human impact. *Freshwat. Biol.* **2008**, *53*, 617–634. [CrossRef]
4. De Castro-Català, N.; Muñoz, I.; Armendáriz, L.; Campos, B.; Barceló, D.; López-Doval, J.; Pérez, S.; Petrovic, M.; Picó, Y.; Riera, J.L. Invertebrate community responses to emerging water pollutants in Iberian river basins. *Sci. Total Environ.* **2015**, *503*, 142–150. [CrossRef] [PubMed]
5. Wallace, J.B.; Webster, J.R. The Role of Macroinvertebrates in Stream Ecosystem Function. *Annu. Rev. Entomol.* **1996**, *41*, 115–139. [CrossRef]
6. Covich, A.P.; Palmer, M.A.; Crowl, T.A. The Role of Benthic Invertebrate Species in Freshwater Ecosystems: Zoobenthic species influence energy flows and nutrient cycling. *BioScience* **1999**, *49*, 119–127. [CrossRef]
7. McLenaghan, N.A.; Tyler, A.C.; Mahl, U.H.; Howarth, R.W.; Marino, R.M. Benthic macroinvertebrate functional diversity regulates nutrient and algal dynamics in a shallow estuary. *Mar. Ecol. Prog. Ser.* **2011**, *426*, 171–184. [CrossRef]
8. Ogbeibu, A.E.; Oribhabor, B.J. Ecological impact of river impoundment using benthic macro-invertebrates as indicators. *Water Res.* **2002**, *36*, 2427–2436. [CrossRef]
9. Arimoro, F.O.; Ikomi, R.B. Ecological integrity of upper Warri River, Niger Delta using aquatic insects as bioindicators. *Ecol. Indic.* **2009**, *9*, 455–461. [CrossRef]
10. Bonada, N.; Prat, N.; Resh, V.H.; Statzner, B. Developments in aquatic insect biomonitoring: A comparative analysis of recent approaches. *Annu. Rev. Entomol.* **2006**, *51*, 495–523. [CrossRef]
11. Birk, S.; Bonne, W.; Borja, A.; Brucet, S.; Courrat, A.; Poikane, S.; Solimini, A.; van de Bund, W.; Zampoukas, N.; Hering, D. Three hundred ways to assess Europe's surface waters: An almost complete overview of biological methods to implement the Water Framework Directive. *Ecol. Indic.* **2012**, *18*, 31–41. [CrossRef]
12. Feld, C.K.; Hering, D. Community structure or function: Effects of environmental stress on benthic macroinvertebrates at different spatial scales. *Freshwat. Biol.* **2007**, *52*, 1380–1399. [CrossRef]
13. Zhang, F.; Wang, J.; Wang, X. Recognizing the Relationship between Spatial Patterns in Water Quality and Land-Use/Cover Types: A Case Study of the Jinghe Oasis in Xinjiang, China. *Water* **2018**, *10*, 646. [CrossRef]
14. Kim, D.-K.; Kaluskar, S.; Mugalingam, S.; Arhonditsis, G.B. Evaluating the relationships between watershed physiography, land use patterns, and phosphorus loading in the Bay of Quinte, Ontario, Canada. *J. Great Lakes Res.* **2016**, *42*, 972–984. [CrossRef]
15. Sponseller, R.A.; Benfield, E.F.; Valett, H.M. Relationships between land use, spatial scale and stream macroinvertebrate communities. *Freshwat. Biol.* **2001**, *46*, 1409–1424. [CrossRef]
16. Kim, D.-K.; Kaluskar, S.; Mugalingam, S.; Blukacz-Richards, A.; Long, T.; Morley, A.; Arhonditsis, G.B. A Bayesian approach for estimating phosphorus export and delivery rates with the SPAtially Referenced Regression On Watershed attributes (SPARROW) model. *Ecol. Inform.* **2017**, *37*, 77–91. [CrossRef]
17. Wellen, C.; Arhonditsis, G.B.; Labencki, T.; Boyd, D. Application of the SPARROW model in watersheds with limited information: A Bayesian assessment of the model uncertainty and the value of additional monitoring. *Hydrol. Process.* **2014**, *28*, 1260–1283. [CrossRef]
18. Park, S.-B.; Lee, S.-K.; Chang, K.-H.; Jeong, K.-S.; Joo, G.-J. The impact of monsoon rainfall (Changma) on the changes of water quality in the lower Nakdong River (Mulgeum). *Korean J. Limnol.* **2002**, *35*, 161–170.
19. Park, J.-S.; Kang, H.-S.; Lee, Y.S.; Kim, M.-K. Changes in the extreme daily rainfall in South Korea. *Int. J. Climatol.* **2011**, *31*, 2290–2299. [CrossRef]
20. Wetzel, R.G.; Likens, G.E. *Limnological Analysis*; Springer: Berlin/Heidelberg, Germany, 1991; p. 429.
21. Quigley, M. *Invertebrates of Streams and Rivers*; Edward A. Ltd.: Colchester, London, UK, 1977; p. 84.
22. Pennak, R.W. *Freshwater Invertebrates of the United States*; John Wiley and Sons, Inc.: Hoboken, MY, USA, 1978; p. 803.
23. Brighnam, A.R.; Brighnam, W.U.; Gnika, A. *Aquatic Insects and Oligochaetea of North and South Carolina*; Midwest Aquatic Enterprises: Seaford, UK, 1982.

24. Yun, I.-B. *Illustrated Encyclopedia of Fauna and Flora of Korea. Aquatic Insects*; Ministry of Education: Seoul, Korea, 1988; Volume 30, p. 840.
25. Merritt, R.W.; Cummins, K.W. *An Introduction to the Aquatic Insects of North America*; Kendall/Hunt Publishing Company: Dubuque, IA, USA, 1996; p. 862.
26. McNaughton, S.J. Relationships among Functional Properties of Californian Grassland. *Nature* **1967**, *216*, 168–169. [CrossRef]
27. Shannon, C.E.; Weaver, W. *The Mathematical Theory of Communication*; The University of Illinois Press: Champaign, IL, USA, 1964; p. 125.
28. Margalef, R. Temporal succession and spatial heterogeneity in phytoplankton. In *Perspectives in Marine Biology*; Buzzati-Traverso, A.A., Ed.; University of California Press: Berkeley, CA, USA, 1958; pp. 323–347.
29. Pielou, E.C. *Ecological Diversity*; Wiley: New York, NY, USA, 1975; p. 165.
30. Kong, D.; Son, S.-H.; Hwang, S.-J.; Won, D.H.; Kim, M.C.; Park, J.H.; Jeon, T.-S.; Lee, J.E.; Kim, J.H.; Kim, J.S.; et al. Development of benthic macroinvertebrates index (BMI) for biological assessment on stream environment. *J. Korean Soc. Water Environ.* **2018**, *34*, 183–201, (Written In Korean).
31. Zelinka, M.; Marvan, P. Zur Präzisierung der biologischen Klassifikation der Reinheit fließender Gewässer. *Arch. Hydrobiol.* **1961**, *57*, 389–407.
32. National Institute of Environmental Research (NIER). *Biomonitoring Survey and Assessment Manual*; NIER: Incheon, Korea, 2017.
33. Lepš, J.; Šmilauer, P. *Multivariate Analysis of Ecological Data Using CANOCO*; Cambridge University Press: Cambridge, UK, 2003; p. 269.
34. Ter Braak, C.J.F.; Smilauer, P. *CANOCO Reference Manual and CanoDraw for Windows User's Guide: Software for Canonical Community Ordination (version 5.0)*; Microcomputer Power: Ithaca, NY, USA, 2012; p. 496.
35. Aschonitis, V.G.; Feld, C.K.; Castaldelli, G.; Turin, P.; Visonà, E.; Fano, E.A. Environmental stressor gradients hierarchically regulate macrozoobenthic community turnover in lotic systems of Northern Italy. *Hydrobiologia* **2016**, *765*, 131–147. [CrossRef]
36. Osborne, J.W. Improving your data transformation: Applying the Box-Cox transformation. *Pract. Assess. Res. Eval.* **2010**, *15*, 1–9.
37. Liu, R.; Wang, Q.; Xu, F.; Men, C.; Guo, L. Impacts of manure application on SWAT model outputs in the Xiangxi River watershed. *J. Hydrol.* **2017**, *555*, 479–488. [CrossRef]
38. Tiessen, K.H.D.; Elliott, J.A.; Yarotski, J.; Lobb, D.A.; Flaten, D.N.; Glozier, N.E. Conventional and conservation tillage: Influence on seasonal runoff, sediment, and nutrient losses in the Canadian Prairies. *J. Environ. Qual.* **2010**, *39*, 964–980. [CrossRef]
39. Ahn, K.S. The water pollution of Yocheon, uppermost stream of the Sumjin River. *J. Korean Earth Sci. Soc.* **2005**, *26*, 821–827.
40. Liang, T.; Wang, S.; Cao, H.; Zhang, C.; Li, H.; Li, H.; Song, W.; Chong, Z. Estimation of ammonia nitrogen load from nonpoint sources in the Xitiao River catchment, China. *J. Environ. Sci.* **2008**, *20*, 1195–1201. [CrossRef]
41. Chen, A.; Lei, B.; Hu, W.; Lu, Y.; Mao, Y.; Duan, Z.; Shi, Z. Characteristics of ammonia volatilization on rice grown under different nitrogen application rates and its quantitative predictions in Erhai Lake Watershed, China. *Nutr. Cycl. Agroecosyst.* **2015**, *101*, 139–152. [CrossRef]
42. Bünemann, E.K.; Heenan, D.P.; Marschner, P.; McNeill, A.M. Long-term effects of crop rotation, stubble management and tillage on soil phosphorus dynamics. *Aust. J. Soil Res.* **2006**, *44*, 611–618. [CrossRef]
43. Easton, Z.M.; Gérard-Marchant, P.; Walter, M.T.; Petrovic, A.M.; Steenhuis, T.S. Identifying dissolved phosphorus source areas and predicting transport from an urban watershed using distributed hydrologic modeling. *Water Resour. Res.* **2007**, *43*. [CrossRef]
44. Ahn, S.R.; Jeong, J.H.; Kim, S.J. Assessing drought threats to agricultural water supplies under climate change by combining the SWAT and MODSIM models for the Geum River basin, South Korea. *Hydrol. Sci. J.* **2016**, *61*, 2740–2753. [CrossRef]
45. Jeong, K.-S.; Hong, D.-G.; Byeon, M.-S.; Jeong, J.-C.; Kim, H.-G.; Kim, D.-K.; Joo, G.-J. Stream modification patterns in a river basin: Field survey and self-organizing map (SOM) application. *Ecol. Inform.* **2010**, *5*, 293–303. [CrossRef]
46. Jäch, M.A.; Balke, M. Global diversity of water beetles (Coleoptera) in freshwater. *Hydrobiologia* **2008**, *595*, 419–442. [CrossRef]

47. Jung, S.W.; Jäch, M.A.; Bae, Y.J. Review of the Korean Elmidae (Coleoptera: Dryopoidea) with descriptions of three new species. *Aquat. Insects* **2014**, *36*, 93–124. [CrossRef]
48. Criado, F.G.; Alaez, M.F. Aquatic Coleoptera (Hydraenidae and Elmidae) as indicators of the chemical characteristics of water in the Orbigo River basin (N-W Spain). *Ann. Limnol. Int. J. Lim.* **1995**, *31*, 185–199. [CrossRef]
49. Wallace, J.B. Food Partitioning in Net-spinning Trichoptera Larvae: *Hydropsyche venularis*, *Cheumatopsyche etrona*, and *Macronema zebratum* (Hydropsychidae). *Ann. Entomol. Soc. Am.* **1975**, *68*, 463–472. [CrossRef]
50. Fuller, R.L.; Mackay, R.J. Effects of food quality on the growth of three *Hydropsyche* species (Trichoptera: Hydropsychidae). *Can. J. Zool.* **1981**, *59*, 1133–1140. [CrossRef]
51. Osborne, L.L.; Herricks, E.E. Microhabitat Characteristics of Hydropsyche (Trichoptera:Hydropsychidae) and the Importance of Body Size. *J. N. Am. Benthol. Soc.* **1987**, *6*, 115–124. [CrossRef]
52. Beketov, M. Different sensitivity of mayflies (Insecta, Ephemeroptera) to ammonia, nitrite and nitrate: Linkage between experimental and observational data. *Hydrobiologia* **2004**, *528*, 209–216. [CrossRef]
53. Lammers-Campbell, R. Ordination of Chironomid (Diptera: Chironomidae) Communities Characterizing Habitats in a Minnesota Peatland. *J. Kans. Entomol. Soc.* **1998**, *71*, 414–425.
54. Principe, R.E.; Boccolini, M.F.; Corigliano, M.C. Structure and Spatial-Temporal Dynamics of Chironomidae Fauna (Diptera) in Upland and Lowland Fluvial Habitats of the Chocancharava River Basin (Argentina). *Int. Rev. Hydrobiol.* **2008**, *93*, 342–357. [CrossRef]
55. Campbell, B.D.; Haro, R.J.; Richardson, W.B. Effects of agricultural land use on chironomid communities: Comparisons among natural wetlands and farm ponds. *Wetlands* **2009**, *29*, 1070–1080. [CrossRef]
56. Corkum, L.D. Responses of chlorophyll-a, organic matter, and macroinvertebrates to nutrient additions in rivers flowing through agricultural and forested land. *Arch. Hydrobiol.* **1996**, *136*, 391–411.
57. Hemphill, N. Competition between two stream dwelling filter-feeders, *Hydropsyche oslari* and *Simulium virgatum*. *Oecologia* **1988**, *77*, 73–80. [CrossRef]
58. Azrina, M.Z.; Yap, C.K.; Rahim Ismail, A.; Ismail, A.; Tan, S.G. Anthropogenic impacts on the distribution and biodiversity of benthic macroinvertebrates and water quality of the Langat River, Peninsular Malaysia. *Ecotoxicol. Environ. Saf.* **2006**, *64*, 337–347. [CrossRef]
59. Scrimgeour, G.J.; Winterbourn, M.J. Effects of floods on epilithon and benthic macroinvertebrate populations in an unstable New Zealand river. *Hydrobiologia* **1989**, *171*, 33–44. [CrossRef]
60. Robinson, C.T.; Uehlinger, U.; Monaghan, M.T. Effects of a multi-year experimental flood regime on macroinvertebrates downstream of a reservoir. *Aquat. Sci.* **2003**, *65*, 210–222. [CrossRef]

Article

Effects of di-(2-ethylhexyl) phthalate on Transcriptional Expression of Cellular Protection-Related *HSP60* and *HSP67B2* Genes in the Mud Crab *Macrophthalmus japonicus*

Kiyun Park [1], Won-Seok Kim [2] and Ihn-Sil Kwak [1,2,*]

[1] Fisheries Science Institute, Chonnam National University, Yeosu 59626, Korea; ecoblue@hotmail.com
[2] Faculty of Marine Technology, Chonnam National University, Yeosu 59626, Korea; csktjr123@gmail.com
* Correspondence: iskwak@chonnam.ac.kr; Tel.: +82-61-6597148; Fax: +82-61-6597149

Received: 10 March 2020; Accepted: 14 April 2020; Published: 16 April 2020

Abstract: Di-2-ethylhexyl phthalate (DEHP) has attracted attention as an emerging dominant phthalate contaminant in marine sediments. *Macrophthalmus japonicus*, an intertidal mud crab, is capable of tolerating variations in water temperature and sudden exposure to toxic substances. To evaluate the potential effects of DEHP toxicity on cellular protection, we characterized the partial open reading frames of the stress-related heat shock protein 60 (*HSP60*) and small heat shock protein 67B2 (*HSP67B2*) genes of *M. japonicus* and further investigated the molecular effects on their expression levels after exposure to DEHP. Putative *HSP60* and small *HSP67B2* proteins had conserved HSP-family protein sequences with different C-terminus motifs. Phylogenetic analysis indicated that *M. japonicus HSP60* (*Mj-HSP60*) and *M. Japonicus HSP67B2* (*Mj-HSP67B2*) clustered closely with *Eriocheir sinensis HSP60* and *Penaeus vannamei HSP67B2*, respectively. The tissue distribution of Heat shock proteins (HSPs) was the highest in the gonad for *Mj-HSP60* and in the hepatopancreas for *Mj-HSP67B2*. The expression of *Mj-HSP60* Messenger Ribonucleic Acid (mRNA) increased significantly at day 1 after exposure to all doses of DEHP, and then decreased in a dose-dependent and exposure time-dependent manner in the gills and hepatopancreas. *Mj-HSP67B2* transcripts were significantly upregulated in both tissues at all doses of DEHP and at all exposure times. These results suggest that cellular immune protection could be disrupted by DEHP toxicity through transcriptional changes to HSPs in crustaceans. Small and large HSPs might be differentially involved in responses against environmental stressors and in detoxification in *M. japonicus* crabs.

Keywords: di(2-ethylhexyl) phthalate (DEHP); crustacean; heat shock proteins (HSPs); gene expression; environmental risk assessment

1. Introduction

Artificial chemical additives have come to the fore as one of the main environmental pollution triggers. Plasticizers, which assign flexibility and durability to plastic, have been heavily utilized, owing to the widespread application of plastic products. As the most common plasticizer, di-2-ethylhexyl phthalate (DEHP) has contributed to the manufacture of flexible products from solid plastics such as polyvinyl chloride [1]. Owing to its widespread use, DEHP is ubiquitously released into the aquatic environment [2,3]. A recent study showed that the main source of DEHP is emissions from household sewage and sludge disposal activities [2]. DEHP is detected at high levels in all sediment samples taken from coastal bays, indicating ubiquitous contamination of the marine environment [3]. DEHP concentrations were found to range from 3020 to 3970 ng/g in sediments from the Kuwait Coast, Pearl River Delta in China, and Kaohsiung Harbor in Taiwan [4–6]. In addition, in the northwestern

Mediterranean Sea, the range of DEHP concentrations was 42–802 ng L^{-1} and 130–924 ng L^{-1} in the surface seawater (depth 0.5 m) and bottom seawater (depth 30 m), respectively [7]. DEHP concentrations were found to range from 62 to 4352 ng L^{-1} from the bottom to the surface seawater of the Bohai Sea and the Yellow Sea, China [8]. DEHP, an endocrine-disrupting chemical (EDC), exhibits a perturbing effect on steroidogenesis activities [9,10].

Macrophthalmus japonicus is one of the main benthic species ubiquitously detected in tidal flats and shows high distribution rates in estuarine regions of Korea and Japan [11,12]. As a main member of the tidal flat food chain, this species contributes to the maintenance of biodiversity in estuarine ecosystems. Because of their dominant distribution, crabs might be a good candidate organism to sense changes in the condition of the surrounding environment, as well as changes involving food reserves, as they have abundant nutrients and are of high economic value in commercial fisheries. However, crab habitats are easily exposed to great hazards, such as plastic waste pollutants and chemicals that are transported into mud flats through rivers or from the ocean. The effects of various stress conditions, such as salinity and heavy metal and biocide contaminants, have been reported following expression analysis of immune-related or stress-related genes in crabs [11,13–16]. A recent study showed the relationships between EDCs and gene expression alterations involving crab innate immune systems [17], but there have been no studies of the relationship between stress-related gene expression and EDC exposure. Despite its biological importance as a nutritional resource, few studies have been conducted on the *M. japonicus* genomic DNA sequence.

Heat shock proteins (HSPs) are ubiquitous proteins secreted in cells after exposure to stressful conditions and are classified into six major groups (*HSP27, HSP60, HSP70, HSP90,* and large HSPs) based on their molecular weights [18,19]. HSPs function as molecular chaperones to prevent the formation of denatured proteins during high temperature stress and exhibit upregulation in their expression patterns under such stress conditions [18,20]. In addition, these stress proteins play an important role in the maintenance of normal polypeptide structures and in the promotion of correct refolding of cellular proteins in response to various external stimuli, such as anoxia, heavy metals, or chemicals, which cause protein denaturation [20–22]. HSPs assist in protecting cellular homeostasis from such stress. *HSP60* is well known as a pro-apoptotic molecule, which induces apoptosis and acts as a chaperone for proteins transcribed from mitochondrial DNA [23–25]. *HSP60* is a highly immunogenic protein, which is implicated in a variety of autoimmune diseases [26,27]. The upregulation of *HSP60* indicates its involvement in crucial functions mediating immune responses in the Chinese mitten crab, *Eriocheir sinensis*, after crustacean pathogen infection [27]. *HSP67B2* was characterized as a Relish-regulated gene in the innate immunity of the Chinese shrimp (*Fenneropenaeus chinensis*) [28]. However, there is limited information about the molecular characterization and expression responses involving the crustacean *HSP67B2*.

In the present study, we identify two stress-related genes, *Mj-HSP60* and *Mj-HSP67B2*, in *M. japonicus* crabs to evaluate the toxic effects of DEHP on cellular immune protection in crustaceans. We investigate the genomic structure, phylogenetic relationships with other homologous HSPs, and transcriptional responses of HSPs under DEHP stress. We seek to provide molecular information regarding the influence of EDCs on stress-related gene expression in *M. japonicus*.

2. Materials and Methods

2.1. Ethical Statement

All experiments involving *M. japonicus* crabs in this study were carried out in accordance with the guidelines and regulations approved by the Institutional Animal Care and Use Committee of Chonnam National University.

2.2. Preparation of *M. japonicus* Individuals

Crabs used in this study were collected from the Yeosu marine products market in Korea. All individuals involved were 3 ± 0.5 cm in shell height, 3.5 ± 0.8 cm in shell width, and 7.5 ± 3.5 g in body weight. We prepared glass tanks (45.7 × 35.6 × 30.5 cm) filled with seawater at 18 °C, with 25% salinity and a photoperiod of 12 h. Crabs were stabilized in glass tanks for 1 day prior to exposure to DEHP solutions. After 1 day, healthy, undamaged crabs were selected for DEHP exposure experiments (below).

2.3. DEHP Exposure Experiments

DEHP solutions were made from a solid compound (99%, Junsei Chemical Co. Ltd., Tokyo, Japan). For preparation of a 10 mg L^{-1} stock solution of DEHP, we dissolved DEHP in 99% acetone at room temperature. This stock solution was diluted with seawater for DEHP solutions with concentrations of 1, 10, and 30 μg L^{-1}. A concentration of <0.5% acetone was used as a solvent control. For the DEHP exposure experiments, a total of 40 crabs were randomly divided into four experimental groups (1, 10, and 30 μg L^{-1} DEHP solutions and solvent control). Ten crabs were placed in each glass tank and exposed to one of the three doses of DEHP over days 1, 4, and 7, respectively. Three individuals were selected for tissue extraction at each time interval from the DEHP treatment and control groups. Food was not provided for the crabs, but seawater with equivalent concentrations of DEHP was added every day during the experiments. The experiments were conducted in triplicate with independent samples.

2.4. Total RNA Extraction and cDNA Synthesis

Crab gill and hepatopancreatic tissues were acquired from the exposure and control groups. Total RNA was extracted using TRIzol reagent (Life Technologies, Rockville, MD, USA) with Recombinant DNase I (Takara, Otsu, Japan) according to the manufacturers' protocols. The concentration of each RNA sample was measured using a Nano-Drop 1000 (Thermo Fisher Scientific, Waltham, MA, USA). RNA integrity was checked by 1% agarose gel electrophoresis. Single-stranded Complementary Deoxyribonucleic Acid (cDNA) synthesis was carried out with 1000 ng of total RNA using an oligo dT primer (50 μM) for reverse transcription in 20 μL reactions (PrimeScript™ 1st strand cDNA synthesis kit, Takara) according to the manufacturer's protocol.

2.5. Gene Expression Analysis Using Quantitative Reverse-Transcription PCR (RT-PCR) Amplification

To confirm the expression patterns of *Mj-HSP60* and *Mj-HSP67B2* in various tissues of *M. japonicus*, and in the control and DEHP-exposed samples, quantitative RT-PCR was carried out on an ExicyclerTM96 instrument (Bioneer, Daejeon, Korea). Each reaction was conducted in a final volume of 20 μL containing 10 μL of Accuprep®2 × Greenstar qPCR Master Mix (Bioneer, Daejeon, Korea), 6 μL of DEPC-treated water, 0.5 μL each of sense primer and antisense primer (10 pM), and 3 μL of 30-fold diluted cDNA sample as a template. Quantitative RT-PCR of two genes was carried out for 40 cycles of 95 °C for 15 s and 60 °C for 45 s using the following primer pairs: *Mj-HSP60* forward 5′-CCCTGAAGGATGAGCTTGAG-3′; *Mj-HSP60* reverse 5′-GCTGGGATGATGGA CTGAAT-3′; *Mj-HSP67B2* forward 5′-GAGCCGCGGTAGATTCTAT G-3′; *Mj-HSP67B2* reverse 5′-CTGGACAAGGAGGGTTTCAA-3′; Glyceraldehyde-3-Phosphate Dehydrogenase (GAPDH) forward 5′-TGCTGATGCACCCATGTTT G-3′; and *GAPDH* reverse 5′-AGGCCCTGGACAATCTCAA AG-3′. Melting curves were determined by increasing the temperature from 68 °C to 94 °C. All samples were amplified in triplicate to ensure reproducibility. The relative expression level of each transcript was determined using *M. japonicus GAPDH* as an internal reference gene and employing the $2^{-\Delta\Delta Ct}$ method [29].

2.6. M. japonicus Hsp Identification and Bioinformatics Analysis

Two HSP genes (*Mj-HSP60* and *Mj-HSP67B2*) were identified by screening a previously generated 454 GS-FLX transcriptome database. Sequences were analyzed based on nucleotide and protein databases using the BLASTN and BLASTX programs (National Center for Biotechnology Information, U.S. National Library of Medicine, Bethesda, MD, USA), respectively [30]. Two domains, the chaperonin-like super family of *Mj-HSP60* and Rhodonase (RHOD) superfamily of *Mj-HSP67B2*, were identified by PROSITE profile analysis [31]. A phylogenetic tree for the two HSPs was generated by the neighbor joining method using Molecular Evolutionary Genetic Analysis (MEGA X, Pennsylvania State University, State College, PA, USA) [32] with 1000 bootstrap replications.

2.7. Statistical Analysis

The Statistical Package for the Social Sciences (SPSS) 12.0 KO (SPSS Inc., Chicago, IL, USA) was used for statistical analysis in this study. Data are presented as the mean ± standard deviation. Two-way analysis of variance was conducted to identify the statistical effects of the exposure period and each DEHP dose on *Mj-HSP60* and *Mj-HSP67B2* mRNA expression. Significant differences were presented as $^{*}P < 0.05$ and $^{**}P < 0.01$.

3. Results

3.1. Characterization of Mj-HSP60 and Mj-HSP67B2 in M. japonicus

We identified two HSP genes (*Mj-HSP60* and *Mj-HSP67B2*) in our 454 GS-FLX transcriptome analysis [33] that were composed of 1360 nucleotides (nt) and 511 nt, which comprised open reading frames encoding 330 and 149 amino acids, respectively (Figures 1A and 2A). *Mj-HSP60* encoded a mature protein of 330 amino acids, 75 bp of 5′ untranslated region (UTR) and 57 bp of 3′ UTR, with a putative methionine initiation codon (ATG) beginning at 58 nt and a stop codon ending at 1224 nt. The SignalP Server (ExPASy) [34] predicted that the first 28 amino acids in the N-terminal region of the polypeptide chain would form a signal peptide sequence. We found that *Mj-HSP60* included a chaperonin-like super family main domain, whereas a RHOD superfamily motif was detected in *Mj-HSP67B2* (Figure 2A). The predicted molecular mass of the deduced amino acid sequence was 61 kDa, with an estimated isoelectric point (pI) of 5.74. *Mj-HSP60* was identified by a BLAST search of the National Center for Biotechnology Information (NCBI) non-redundant (nr) database. To understand the evolutionary position of the *Mj-HSP60*, we undertook phylogenetic analysis using another 11 species of crustaceans. As shown in Figure 1B, the phylogenetic tree consisted of two clades involving 12 crustacean species. The *Mj-HSP60* formed one main clade with other crabs (*Eriocheir sinensis*, *Scylla paramamosain*, and *Portunus trituberculatus*) and crayfish (*Cherax cainii*, *Cherax quadricarinatus*, and *Cherax destructor*). The other clade was composed of shrimp species (*Macrobrachium nipponense*, *Macrobrachium rosenbergii*, *Penaeus japonicus*, *Penaeus monodon*, and *Penaeus vannamei*). For clear annotation of *Mj-HSP67B2*, we examined the RHOD superfamily domain sequence (98 amino acids) using BLASTN searches of the nr database to detect sequences of other species with high similarity. We carried out pairwise alignment of *Mj-HSP67B2* using EMBOSS alignment (EMBL-EBI, Cambridgeshire, UK) [35] with sequences identified in BLAST searches. The results showed 35.9–72.8% sequence identity, 54.4–81.6% similarity, and 4.9–10.5% gap percentage when compared with *HSP67B2* from other species (Table 1). The *Mj-HSP67B2* sequence revealed considerable identity (72.8%), similarity (81.6%), and gap percentage (4.9%) with *Penaeus vannamei HSP67B2*. In addition, phylogenetic analysis of the *Mj-HSP67B2* was carried out using data from various arthropod species, owing to deficient genomic information regarding the *HSP67B2* in crustaceans (Figure 2B). The results showed that the two main clades were divided into Crustacea and Insecta, including mosquito and fly species. The *Mj-HSP67B2* showed the closest phylogenetic relationship to *Penaeus vannamei HSP67B2*. Given these results from analysis of phylogenetic and pairwise sequence alignment comparisons, our transcript sequence from the transcriptome database was identified as *Mj-HSP67B2*.

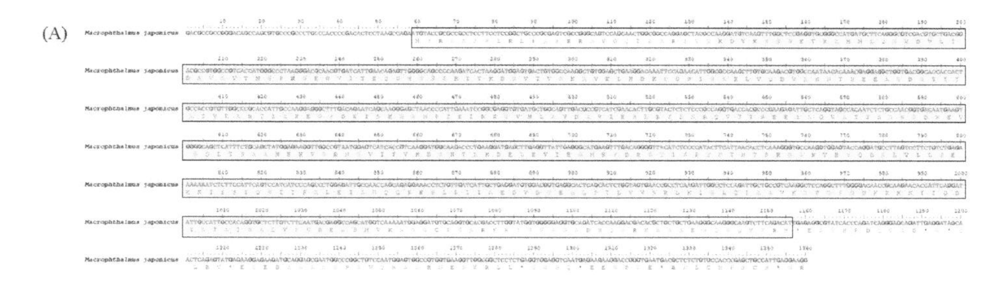

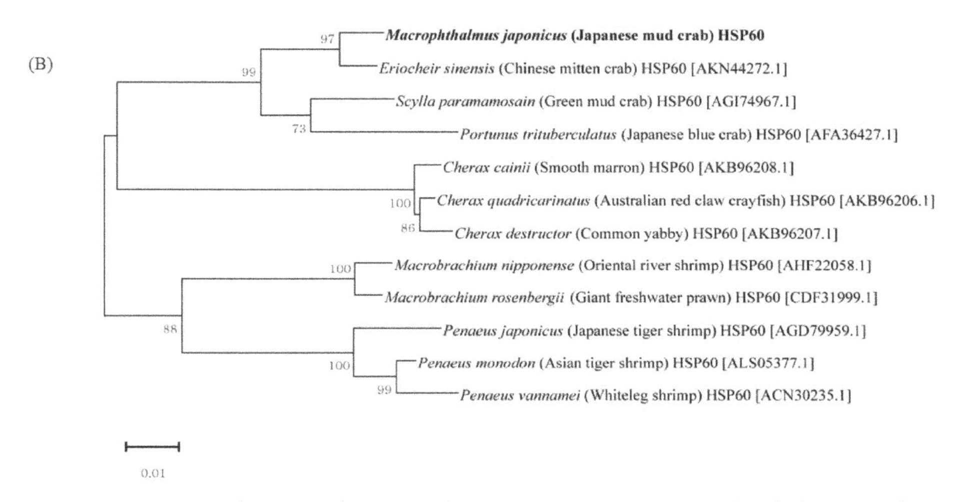

Figure 1. Genomic information of *Macrophthalmus japonicus HSP60* sequences identified in this study. (**A**) *Mj-HSP60* structure was represented using the BioEdit program (North Carolina State University, Raleigh, NC, USA). The open reading frame (ORF) of *Mj-HSP60* was predicted using the ExPASy tool and is shown as a black box. The yellow box indicates the chaperonin-like super family domain. (**B**) Phylogenetic analysis of *Mj-HSP60* with known *HSP60* sequences from 11 Crustacean species. The phylogenetic tree is based on amino acid sequences translated from *Mj-HSP60* ORF by the neighbor joining method (bootstrap value 1000) using MEGA X software. The numbers at the nodes represent the bootstrap majority consensus values for 1000 replicates. GenBank accession numbers are shown with scientific and common names of each species.

3.2. Expression Analysis of Mj-HSP60 and Mj-HSP67B2 in Various Tissues of M. japonicus

To better understand the expression patterns of *Mj-HSP60* and *Mj-HSP67B2*, quantitative RT-PCR was carried out for six tissue sources (gill, hepatopancreas, muscle, gonad, heart, and stomach) of *M. japonicus*. The highest level of *Mj-HSP60* expression was found in the gonad, while *Mj-HSP67B2* was predominantly expressed in the hepatopancreas (Figure 3). In the gonad, *Mj-HSP60* was expressed 3.7-fold higher than *Mj-HSP67B2*. In contrast, *Mj-HSP67B2* exhibited a higher expression level than *Mj-HSP60* in the gills (1.7-fold) and hepatopancreas (3.1-fold). Relatively low levels of *Mj-HSP60* and *Mj-HSP67B2* expression were observed in the muscle, heart, and stomach tissues.

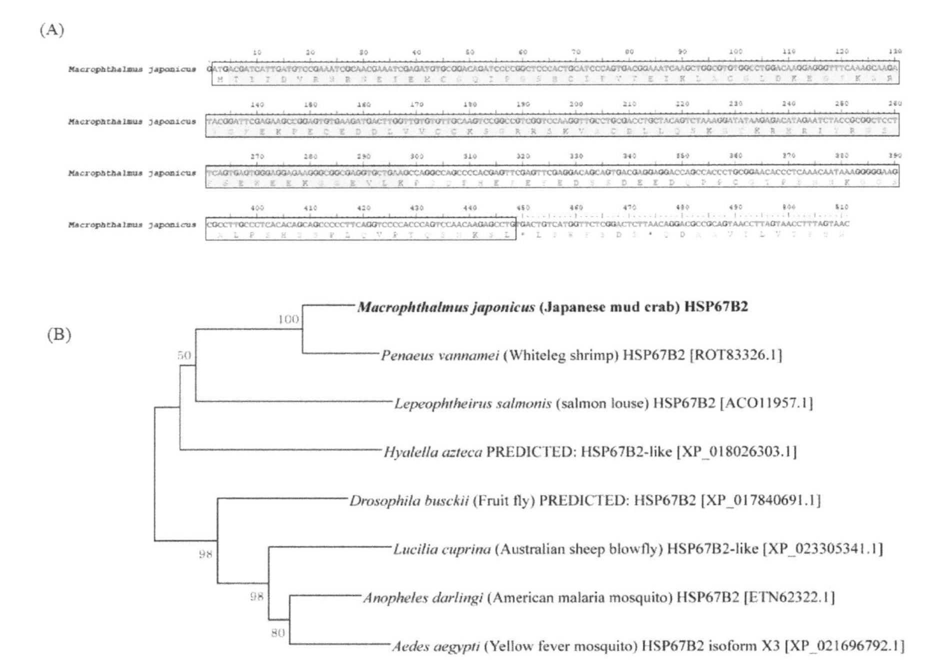

Figure 2. Sequence information for *Macrophthalmus japonicus HSP67B2* identified in this study. (**A**) An open reading frame (ORF) of *Mj-HSP67B2* was predicted using the ExPASy tool and is represented by a black box. The yellow box indicates a RHOD superfamily domain. (**B**) Phylogenetic analysis of *Mj-HSP67B2* with known *HSP67B2* sequences from seven Arthropoda species. The phylogenetic tree is based on amino acid sequences translated from *Mj-HSP67B2* ORF by the neighbor joining method (bootstrap value 1000) using MEGA X software. The numbers at the nodes represents the bootstrap majority consensus values for 1000 replicates. GenBank accession numbers are shown with the scientific and common names of each species.

Table 1. Percentage identity, similarity, and gaps involving *Macrophthalmus japonicus* HSP67B2 and HSP67B2 homologs from other species at the amino acid level

Species	Gene Name	Accession Number	RHOD Superfamily Domain Length	Identity (%)	Similarity (%)	Gap (%)
Macrophthalmus japonicus	Heat Shock protein 67B2		98			
Penaeus vannamei	Heat Shock protein 67B2	ROT83326.1	103	72.8	81.6	4.9
Lepeophtheirus salmonis	Heat Shock protein 67B2	ACO11957.1	106	43.4	60.4	7.5
Lucilia cuprina	Heat Shock protein 67B2-like	XP_023305341.1	101	40.0	60.0	10.5
Hyalella azteca	PREDICTED: heat shock protein 67B2-like	XP_018026303.1	106	44.3	58.5	7.5
Drosophila busckii	PREDICTED: heat shock protein 67B2	XP_017840691.1	99	39.8	58.3	8.7
Anopheles darlingi	Heat Shock protein 67B2	ETN62322.1	103	39.8	56.3	4.9
Aedes aegypti	Heat Shock protein 67B2 isoform X3	XP_021696792.1	99	35.9	54.4	8.7

Pairwise identity percentage was calculated using the EMBOSS alignment program.

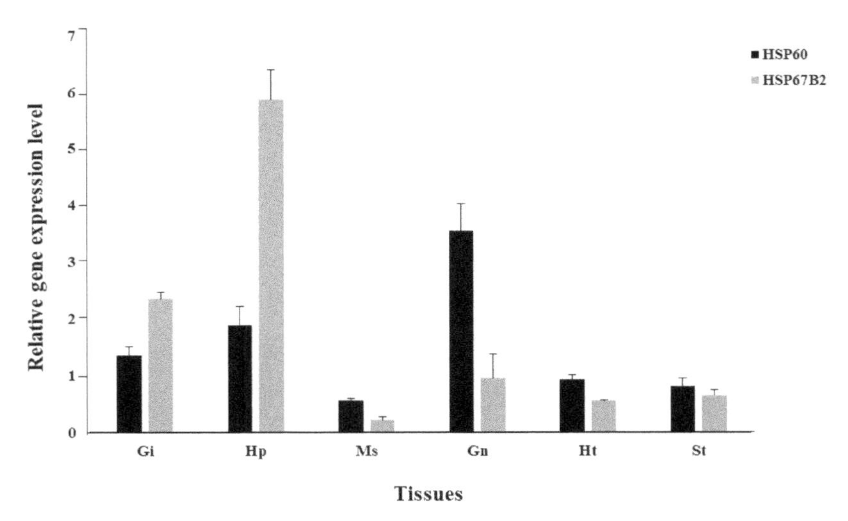

Figure 3. Relative mRNA expression levels of *HSP60* and *HSP67B2* in various *Macrophthalmus japonicus* tissues. Six tissues were used in this experiment. Quantitative reverse-transcription (RT)-PCR was conducted in triplicate. Bars indicate the standard deviation of the mean. mRNA expression was normalized against *GAPDH*. Abbreviations: Gill (Gi), Hepatopancreas (Hp), Muscle (Ms), Gonad (Gn), Heart (Ht), and Stomach (St).

3.3. *M. japonicus Mj-HSP60 Expression Changes after DEHP Exposure*

To confirm the effects of DEHP exposure on *Mj-HSP60* expression, we conducted quantitative RT-PCR analysis using mRNA acquired from the gill and hepatopancreas samples after exposure to DEHP for 1, 4, and 7 days. *Mj-HSP60* was expressed approximately 8.2-fold higher after exposure to 1 μg L^{-1} DEHP ($P < 0.01$), 3.2-fold higher for 10 μg L^{-1} ($P < 0.05$), and 9.4-fold higher for 30 μg L^{-1} ($P < 0.01$) in the gill tissue on day 1 (Figure 4A). With the passage of time, expression levels gradually decreased in all DEHP concentration groups. By day 4, for the 10 and 30 μg L^{-1} treatment groups, expression levels were restored to control levels. By day 7, *Mj-HSP60* expression levels were lower than those of the control. In particular, sharp decreases in expression levels were found in 10 μg L^{-1} (0.3-fold) and 30 μg L^{-1} (0.21-fold) ($P < 0.05$) groups. In the hepatopancreatic tissue, expression levels of *Mj-HSP60* exhibited an overall increased pattern compared to the expression levels in the controls on day 1 (Figure 4B). Expression levels significantly increased by 2.4-fold for 1 μg L^{-1}, 2.6-fold for 10 μg L^{-1}, and 2.9-fold for 30 μg L^{-1} DEHP ($P < 0.05$). By days 4 and 7, *Mj-HSP60* expression levels returned to control levels for the 1 μg L^{-1} group. In the 10 μg L^{-1} DEHP group, *Mj-HSP60* expression decreased to <0.5-fold on day 4, and then recovered slightly toward that of control levels by day 7.

3.4. *Variation in Expression of Mj-HSP67B2 after DEHP Exposure in M. japonicus*

Expression of *Mj-HSP67B2* consistently increased in the gill and hepatopancreatic tissues for 4 days after DEHP exposure at all concentrations (Figure 5). After a peak in expression at day 4, *Mj-HSP67B2* levels somewhat decreased. These *Mj-HSP67B2* expression patterns were found in the two tissues, regardless of DEHP exposure concentration. Although expression levels of *Mj-HSP67B2* decreased after day 4, the expression was still maintained in the gill tissue at higher levels than those of the controls for all concentration groups, except on day 7 (0.86-fold) for the 1 μg L^{-1} group (Figure 5A). Similar changes in *Mj-HSP67B2* expression levels were noted in the hepatopancreas tissue. *Mj-HSP67B2* was strongly overexpressed for 4 days in response to exposure to all concentrations of DEHP ($P < 0.05$), and its expression levels displayed dose-dependent and time-dependent increases for

4 days (Fig. 5B). The highest expression levels were noted on day 4 in each DEHP concentration group (3.9-fold for 1 μg L^{-1} ($P < 0.05$), 5.48-fold for 10 μg L^{-1} ($P < 0.01$), and 5.88-fold for 30 μg L^{-1} ($P < 0.01$).

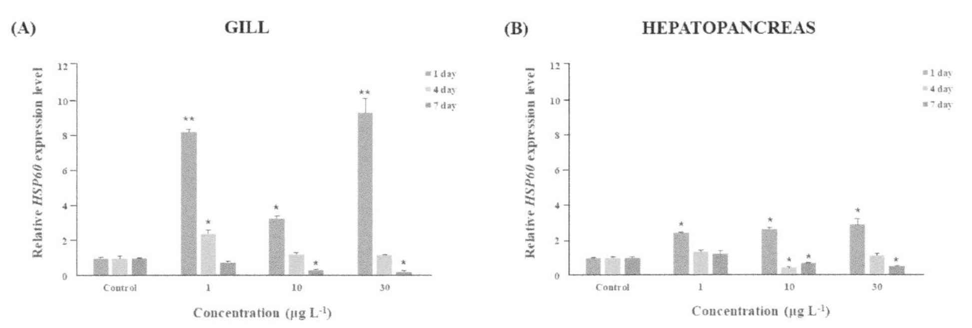

Figure 4. Expression analysis of *HSP60* in the (**A**) gill and (**B**) hepatopancreas of *Macrophthalmus japonicus* exposed to 1, 10, and 30 μg L^{-1} DEHP after 1, 4, and 7 days. Values were normalized against *GAPDH*. Bars indicate the standard deviation of the mean. Statistically significant differences are represented by asterisks as $^{*}P < 0.05$ and $^{**}P < 0.01$, compared to controls (control ratio value = 1).

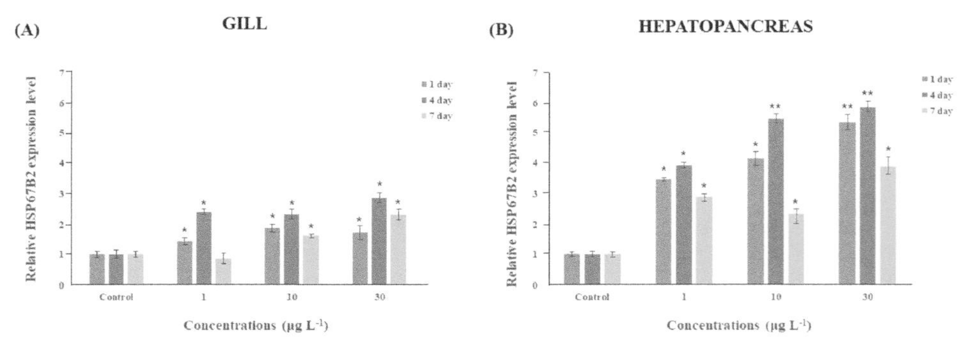

Figure 5. Expression analysis of *HSP67B2* in the (**A**) gill and (**B**) hepatopancreas of *Macrophthalmus japonicus* exposed to 1, 10, and 30 μg L^{-1} DEHP for 1, 4, and 7 days. The values were normalized against *GAPDH*. Bars indicate the standard deviation of the mean. Statistically significant differences are represented by asterisks as $^{*}P < 0.05$ and $^{**}P < 0.01$ as compared to controls (control ratio value = 1).

4. Discussion

Cellular responses to stressors are an evolutionary, ubiquitous, and essential mechanism for cell survival. HSPs are known as extrinsic chaperons that are involved in certain cellular processes, such as germ cell differentiation, reproduction, development, thermoprotection, mammalian autoimmune defense, and toxic stress responses, and they have even been regarded as a potential marker of environmental stress [36–42]. HSPs are found in all eukaryotes and are identified based on their size, molecular weight, and functions. *HSP60*, *HSP70* and *HSP90* are highly conserved genes and are stress-inducible and multigenic [43]. It has been observed that the *HSP60* and *HSP70* family members play significant roles in cell survival, stress, and thermal tolerance in response to various heat shocks [44].

Here, we studied two stress-related genes, *Mj-HSP60* and *Mj-HSP67B2*, and conducted expression analysis in different tissues of *M. japonicus* after treatment with the xenobiotic DEHP. *Mj-HSP60* and *Mj-HSP67B2* were highly expressed in the gonad and hepatopancreas, respectively. In addition, these molecules are moderately expressed in the gills, muscle, heart, and stomach. Our findings are consistent with the results of an earlier study showing that the hepatopancreas is the main source of immune molecules in crustaceans [45]. The hepatopancreas acts as an essential metabolic center in crustaceans and performs versatile roles in defense systems, detoxification, reactive oxygen species production, digestion, absorption, and nutrient secretion. Owing to the critical importance of the hepatopancreas in detoxification and immunological activities, it is highly sensitive to xenobiotic exposure. Similarly, increased upregulation of *HSP90* was noted in the hepatopancreas of *P. monodon* [46]. In addition, three HSPs, namely *MrHSP60*, *MrHSP70* and *MrHSP90*, are constitutively expressed in *M. rosenbergii* during pathogenic infections involving different tissues [47]. Related results were obtained in the Pacific oyster *Crassostrea gigas*, which exhibits highly upregulated *HSP70* expression in the gill tissue after exposure to Cu^{2+} [48]. DEHP has been shown to alter the expression of HSPs in *Chironomus riparius* [49,50]. In this species, *HSP40* and *HSP90* mRNA expression levels increased under various DEHP concentrations for 24 h, which caused morphological deformities [49]. In addition, *HSP70* showed increased expression when treated with low doses of DEHP. Overall, our results indicated that two HSPs, *Mj-HSP60* and *Mj-HSP67B2*, in *M. japonicus* are constitutively expressed, owing to DEHP exposure at day 1. Hence, these molecules can be considered as upregulated responses of xenobiotic levels for the early exposure time in *M. japonicus* crabs. However, at long-term exposure for 7 days, there are different expression patterns between the *Mj-HSP60* and the *Mj-HSP67B2* transcripts. The *Mj-HSP60* expression was downregulated in most crabs after 7 days of DEHP exposure due to reducing cellular immune protection, although expressions of the detoxifying *Mj-HSP67B2* gene [51] were continuously upregulated in DEHP-treated groups compared to the control. *HSP67B2* is significant both in detoxification and in anti-oxidative stress systems, as well as immune protection [26,27,51]. For instance, in *P. trituberculatus*, an important marine and aquaculture species, *Mj-HSP60* displays differential expression patterns in response to environmental salinity stress and exhibits upregulation in the gills [52].

Likewise, *L. vannamei HSP60* mRNA is regulated between 4 and 32 h after the injection of bacteria [53]. *HSP70* is upregulated 24 h after copper exposure in the zebra mussel *Dreissena polymorpha* and midge larvae *Chironomus tentans* [54,55]. In addition, *HSP70* expression is dramatically induced, owing to microbial pathogens in the Chinese shrimp *Fenneropenaeus chinensis* [56]. However, little is known regarding the response of *HSP60* to xenobiotics and stresses in invertebrates such as the sea anemone (*Anemonia viridis*) [29], *D. polymorpha* [54], and the white shrimp (*Litopenaeus vannamei*) [57]. The limited study reported that HSP67B2 acts like a rhodanese homolog with a single RHOD domain, is characterized from the housefly M. domestica, and plays potential roles under oxidative stress conditions [57]. *M. domestica*, and plays potential roles under oxidative stress conditions [51]. In crustaceans, *HSP* expression studies have been conducted on the Asian paddle crab *Charybdis japonica*, with exposure to EDCs (bisphenol A and 4-nonylphenol) [16,58]. To date, this is the first nucleotide and protein sequence information reported regarding *Mj-HSP60* and *Mj-HSP67B2* in the crab species *M. japonicus*. Our gene expression results revealed the potential involvement of the two HSPs in the immune system of crabs. This study highlights the potential importance of these molecules in crustaceans, protecting cells against pathogens as well as in severe cellular and environmental stress conditions.

Author Contributions: Conceptualization, K.P., W.-S.K. and I.-S.K; methodology, K.P., W.-S.K. and I.-S.K; formal analysis, K.P., W.-S.K. and I.-S.K; investigation, K.P., W.-S.K. and I.-S.K; resources, K.P., W.-S.K. and I.-S.K; writing—original draft preparation, K.P., W.-S.K. and I.-S.K; supervision, K.P., W.-S.K. and I.-S.K; project administration, I.S.K; funding acquisition, K.P., I.-S.K. All authors have read and agreed to the published version of the manuscript.

Funding: This study was supported by the National Research Foundation of Korea, South Korea, which is funded by the Korean Government [NRF-2018-R1A6A1A-03024314] and [NRF-2019-R1I1A1A-01056855].

Conflicts of Interest: The authors declare that they have no conflicts of interest.

References

1. Park, K.; Kwak, I.S. Molecular effects of endocrine-disrupting chemicals on the *Chironomus riparius* estrogen-related receptor gene. *Chemosphere* **2010**, *79*, 934–941. [CrossRef]

2. Lee, Y.S.; Lee, S.; Lim, J.E.; Moon, H.B. Occurrence and emission of phthalates and non-phthalate plasticizers in sludge from wastewater treatment plants in Korea. *Sci. Total Environ.* **2019**, *692*, 354–360. [CrossRef]
3. Kim, S.; Lee, Y.S.; Moon, H.B. Occurrence, distribution, and sources of phthalates and non-phthalate plasticizers in sediment from semi-enclosed bays of Korea. *Mar. Pollut. Bull.* **2020**, *151*, 110824. [CrossRef] [PubMed]
4. Li, X.; Yin, P.; Zhao, L. Phthalate esters in water and surface sediments of the Pearl River estuary: Distribution, ecological, and human health risks. *Environ. Sci. Pollut. Res.* **2016**, *23*, 19341–19349. [CrossRef] [PubMed]
5. Saeed, T.; Al-Jandal, N.; Abusam, A.; Taqi, H.; Al-Khabbaz, A.; Zafar, J. Sources and levels of endocrine disrupting compounds (EDCs) in Kuwait's coastal areas. *Mar. Pollut. Bull.* **2017**, *118*, 407–412. [CrossRef] [PubMed]
6. Chen, C.F.; Chen, C.W.; Ju, Y.R.; Dong, C.D. Determination and assessment of phthalate esters content in sediments from Kaohsiung Harbor, Taiwan. *Mar. Pollut. Bull.* **2017**, *124*, 767–774. [CrossRef]
7. Paluselli, A.; Fauvelle, V.; Schmidt, N.; Galgani, F.; Net, S.; Sempéré, R. Distribution of phthalates in Marseille Bay (NW Mediterranean Sea). *Sci. Total Environ.* **2018**, *621*, 578–587. [CrossRef]
8. Zhang, Z.M.; Zhang, H.H.; Zou, Y.W.; Yang, G.P. Distribution and ecotoxicological state of phthalate esters in the sea-surface microlayer, seawater and sediment of the Bohai Sea and the Yellow Sea. *Environ. Pollut.* **2018**, *240*, 235–247. [CrossRef]
9. Park, J.; Park, C.; Gye, M.C.; Lee, Y. Assessment of endocrine-disrupting activities of alternative chemicals for bis (2-ethylhexyl)phthalate. *Environ. Res.* **2019**, *172*, 10–17. [CrossRef]
10. Park, K.; Jo, H.; Kim, D.K.; Kwak, I.S. Environmental pollutants impair transcriptional regulation of the vitellogenin gene in the burrowing mud crab (*Macrophthalmus japonicus*). *Appl. Sci.* **2019**, *9*, 1401. [CrossRef]
11. Park, K.; Nikapitiya, C.; Kim, W.S.; Kwak, T.S.; Kwak, I.S. Changes of exoskeleton surface roughness and expression of crucial participation genes for chitin formation and digestion in the mud crab (*Macrophthalmus japonicus*) following the antifouling biocide irgarol. *Ecotoxicol. Environ. Saf. 2016. 132*, 186–195. [CrossRef]
12. Kitaura, J.; Nishida, M.; Wada, K. Genetic and behavioral diversity in the *Macrophthalmus japonicus* species complex (Crustacea: Brachyura: Ocypodidae). *Mar. Biol.* **2002**, *140*, 1–8.
13. Park, K.; Kwak, T.S.; Kim, W.S.; Kwak, I.S. Changes in exoskeleton surface roughness and expression of chitinase genes in mud crab *Macrophthalmus japonicus* following heavy metal differences of estuary. *Mar. Pollut. Bull.* **2019**, *138*, 11–18. [CrossRef] [PubMed]
14. Nikapitiya, C.; Kim, W.S.; Park, K.; Kim, J.; Lee, M.O.; Kwak, I.S. Chitinase gene responses and tissue sensitivity in an intertidal mud crab (*Macrophthalmus japonicus*) following low or high salinity stress. *Cell Stress Chaperones.* **2015**, *20*, 517–526. [CrossRef] [PubMed]
15. Nikapitiya, C.; Kim, W.S.; Park, K.; Kwak, I.S. Identification of potential markers and sensitive tissues for low or high salinity stress in an intertidal mud crab (*Macrophthalmus japonicus*). *Fish Shellfish Immunol.* **2014**, *41*, 407–416. [CrossRef] [PubMed]
16. Park, K.; Kwak, I.S. Characterize and gene expression of heat shock protein 90 in marine crab *Charybdis japonica* following bisphenol A and 4-nonylphenol exposures. *Environ. Health Toxicol.* **2014**, *29*, e2014002. [CrossRef]
17. Park, K.; Kim, W.S.; Kwak, I.S. Endocrine-disrupting chemicals impair the innate immune prophenoloxidase system in the intertidal mud crab, *Macrophthalmus japonicus*. *Fish Shellfish Immunol.* **2019**, *87*, 322–332. [CrossRef]
18. Díaz, F.; Orobio, R.F.; Chavarriaga, P.; Toro-Perea, N. Differential expression patterns among heat-shock protein genes and thermal responses in the whitefly *Bemisia tabaci* (MEAM 1). *J. Therm. Biol.* **2015**, *52*, 199–207. [CrossRef]
19. Wu, J.; Liu, T.; Rios, Z.; Mei, Q.; Lin, X.; Cao, S. Heat Shock Proteins and Cancer. *Trends Pharmacol. Sci.* **2017**, *38*, 226–256. [CrossRef]
20. Huang, L.H.; Le, K. Cloning and interspecific altered expression of heat shock protein genes in two leafminer species in response to thermal stress. *Insect Mol. Biol.* **2007**, *16*, 491–500. [CrossRef]
21. Liu, T.; Daniels, C.K.; Cao, S. Comprehensive review on the HSC70 functions, interactions with related molecules and involvement in clinical diseases and therapeutic potential. *Pharmacol. Ther.* **2012**, *136*, 354–374. [CrossRef] [PubMed]
22. Macario, A.J.; Conway de Macario, E. Molecular chaperones: Multiple functions, pathologies, and potential applications. *Front. Biosci.* **2007**, *12*, 2588–2600. [CrossRef] [PubMed]

23. Macario, A.J.; Macario, C.E. Sick chaperones, cellular stress, and disease. *N. Engl. J. Med.* **2005**, *353*, 1489–1501. [CrossRef]
24. Chang, H.C.; Tang, Y.C.; Hayer-Hartl, M.; Hartl, U.C. SnapShot: Molecular chaperones. Part I. *Cell* **2007**, *128*, 212. [CrossRef] [PubMed]
25. Tang, Y.C.; Chang, H.C.; Hayer-Hartl, M.; Hartl, U.C. SnapShot: Molecular chaperones. Part II. *Cell* **2007**, *128*, 412. [CrossRef]
26. Xu, X.Y.; Shen, Y.B.; Fu, J.J.; Liu, F.; Guo, S.Z.; Yang, X.M.; Li, J.L. Molecular cloning, characterization and expression patterns of HSP60 in the grass carp (*Ctenopharyngodon idella*). *Fish Shellfish Immunol.* **2011**, *31*, 864–870. [CrossRef]
27. Ning, M.X.; Xiu, Y.J.; Bi, J.X.; Liu, Y.H.; Hou, L.B.; Ding, Z.F.; Gu, W.; Wang, W.; Meng, Q.G. Interaction of heat shock protein 60 (HSP60) with microRNA in Chinese mitten crab during *Spiroplasma eriocheiris* infection. *Dis. Aquat. Organ.* **2017**, *125*, 207–215. [CrossRef]
28. Wang, D.; Li, S.; Li, F. Screening of genes regulated by Relish in Chinese shrimp *Fenneropenaeus chinensis*. *Dev. Comp. Immunol.* **2013**, *41*, 209–216. [CrossRef]
29. Livak, K.J.; Schmittgen, T.D. Analysis of relative gene expression data using real time quantitative PCR and the $2^{-\Delta\Delta CT}$ method. *Methods* **2001**, *25*, 402–408. [CrossRef]
30. Altschul, S.F.; Gish, W.; Miller, W.; Myers, E.W.; Lipman, D.J. Basic local alignment search tool. *J. Mol. Biol.* **1990**, *215*, 403–410. [CrossRef]
31. Bairoch, A.; Bucher, P.; Hofmann, K. The PROSITE database, its status in 1997. *Nucleic Acids Res.* **1997**, *25*, 217–221. [CrossRef]
32. Kumar, S.; Stecher, G.; Knyaz, C.; Tamura, K. MEGA X: Molecular evolutionary genetics analysis across computing platforms. *Mol. Biol. Evol.* **2018**, *35*, 1547–1549. [CrossRef]
33. Park, K.; Nikapitiya, C.; Kwak, I.S. Identification and expression of proteolysis response genes for *Macrophthalmus japonicus* exposure to irgarol toxicity. *Ann. Limnol. Int. J. Limnol.* **2016**, *52*, 65–74. [CrossRef]
34. Gasteiger, E.; Gattiker, A.; Hoogland, C.; Ivanyi, I.; Appel, R.D.; Bairoch, A. ExPASy: The proteomics server for in-depth protein knowledge and analysis. *Nucleic Acids Res.* **2003**, *31*, 3784–3788. [CrossRef] [PubMed]
35. Madeira, F.; Park, Y.M.; Lee, J.; Buso, N.; Gur, T.; Madhusoodanan, N.; Basutkar, P.; Tivey, A.R.N.; Potter, S.C.; Finn, R.D.; et al. The EMBL-EBI search and sequence analysis tools APIs in 2019. *Nucleic Acids Res.* **2019**, *47*, W636–W641. [CrossRef] [PubMed]
36. Kozlova, T.; Perezgasga, T.; Reynaud, E.; Zurita, M. The Drosophila melanogaster homologue of the hsp60 gene is encoded by the essential locus 1 (1)10Ac and is differentially expressed during fly development. *Dev. Genes Evol.* **1997**, *207*, 253–263. [CrossRef] [PubMed]
37. Meinhardt, A.; Wilhem, B.; Seitz, J. Expression of mitochondrial marker proteins during spermatogenesis. *Hum. Reprod. Update* **1999**, *5*, 108–119. [CrossRef]
38. Timakov, B.; Zhang, P. The hsp60B gene in *Drosophila melanogaster* is essential for the spermatid individualization process. *Cell Stress Chaperones* **2001**, *6*, 71–77. [CrossRef]
39. Choresh, O.; Ron, E.; Loya, Y. The 60-kDa heat shock protein (HSP60) of the sea anemone *Anemonia viridis*: A potential early warning system for monitoring environmental changes. *Mar. Biotechnol.* **2001**, *3*, 501–508. [CrossRef]
40. Kammenga, J.E.; Arts, M.S.J.; Oude-Breuil, W.J.M. HSP60 as a potential biomarker of toxic stress in the Nematode *Plectus acuminatus*. *Arch. Environ. Contam. Toxicol.* **1998**, *34*, 253–258. [CrossRef]
41. Chen, Z.; Christina, C.C.H.; Zhang, J.; Cao, L.; Chen, L.; Zhou, L.; Jin, Y.; Ye, H.; Deng, C.; Dai, Z.; et al. Transcriptomic and genomic evolution under constant cold in *Antarctic notothenioid* fish. *Proc. Natl. Acad. Sci. USA* **2008**, *105*, 12944–12949. [CrossRef] [PubMed]
42. Vabulas, R.M.; Ahmad-Nejad, P.; Da Costa, C.; Miethke, T.; Kirschning, C.J.; Haucker, H.; Wagner, H. Endocytosed HSP60s use toll-like receptor 2 (TLR2) and TLR4 to activate the toll/interleukin-1 receptor signaling pathway in innate immune cells. *J. Biol. Chem.* **2001**, *276*, 31332–31339. [CrossRef] [PubMed]
43. Lindquist, S.; Craig, E.A. The heat-shock proteins. *Annu. Rev. Genet.* **1988**, *22*, 631–677. [CrossRef]
44. Parsell, D.A.; Lindquist, S. The function of heat-shock proteins in stress tolerance: Degradation and reactivation of damaged proteins. *Annu. Rev. Genet.* **1993**, *27*, 437–496. [CrossRef] [PubMed]
45. Brunet, M.; Arnaud, J.; Mazz, J. Gut structure and digestive cellular process in marine crustaceans. *Oceanogra. Mar. Biol.* **1990**, *32*, 335–367.

46. Rungrassamee, W.; Leelatanawit, R.; Jiravanichpaisal, P.; Klinbunga, S.; Karoonuthaisiri, N. Expression and distribution of three heat shock protein genes under heat shock stress and under exposure to Vibrio harveyi in *Penaeus monodon*. *Dev. Comp. Immunol.* **2010**, *34*, 1082–1089. [CrossRef]
47. Chaurasia, M.K.; Nizam, F.; Ravichandran, G.; Arasu, M.V.; Al-Dhabi, N.A.; Arshad, A.; Elumalai, P.; Arockiaraj, J. Molecular importance of prawn large heat shock proteins 60, 70 and 90. 2016. *Fish Shellfish Immunol.* **2016**, *48*, 228–238. [CrossRef]
48. Luan, W.; Li, F.; Zhang, J.; Wen, R.; Li, Y.; Xiang, J. Identification of a novel inducible cytosolic Hsp70 gene in Chinese shrimp *Fenneropenaeus chinensis* and comparison of its expression with the cognate Hsc70 under different stresses. *Cell Stress Chaperones* **2010**, *15*, 83–93. [CrossRef]
49. Park, K.; Kwak, I.S. Characterization of heat shock protein 40 and 90 in *Chironomus riparius* larvae: Effects of di (2-ethylhexyl) phthalate exposure on gene expressions and mouthpart deformities. *Chemosphere* **2008**, *74*, 89–95. [CrossRef]
50. Morales, M.; Planelló, R.; Martínez-Paz, P.; Herrero, O.; Cortés, E.; Martínez-Guitarte, J.L.; Morcillo, G. Characterization of Hsp70 gene in *Chironomus riparius*: Expression in response to endocrine disrupting pollutants as a marker of ecotoxicological stress. *Comp. Biochem. Physiol. C Toxicol. Pharmacol.* **2011**, *153*, 150–158. [CrossRef]
51. Tang, T.; Sun, H.; Li, Y.; Chen, P.; Liu, F. MdRDH1, a HSP67B2-like rhodanese homologue plays a positive role in maintaining redox balance in *Musca domestica*. *Mol. Immunol.* **2019**, *107*, 115–122. [CrossRef]
52. Xu, Q.; Qin, Y. Molecular cloning of heat shock protein 60 (PtHSP60) from *Portunus trituberculatus* and its expression response to salinity stress. *Cell Stress Chaperones* **2012**, *17*, 589–601. [CrossRef] [PubMed]
53. Qian, D.; Shi, Z.; Zhang, S.; Cao, Z.; Liu, W.; Li, L. Extra small virus-like particles (XSV) and nodavirus associated with whitish muscle disease in the giant freshwater prawn, *Macrobrachium rosenbergii*. *J. Fish Dis.* **2003**, *26*, 521–527. [CrossRef] [PubMed]
54. Clayton, M.E.; Steinmann, R.; Fent, K. Different expression patterns of heat shock proteins hsp 60 and hsp 70 in zebra mussels (*Dreissena polymorpha*) exposed to copper and tributyltin. *Aquat. Toxicol.* **2000**, *47*, 213–226. [CrossRef]
55. Karouna-Renier, N.K.; Zehr, J.P. Short-term exposures to chronically toxic copperconcentrations induce HSP70 proteins in midge larvae (*Chironomus tentans*). *Sci. Total Environ.* **2003**, *312*, 267–272. [CrossRef]
56. Li, F.H.; Luan, W.; Zhang, C.S.; Zhang, J.Q.; Wang, B.; Xie, Y.S. Cloning of cytoplasmic heat shock protein 90 (FcHSP90) from *Fenneropenaeus chinensis* and its expression response to heat shock and hypoxia. *Cell Stress Chaperones* **2009**, *14*, 161–172. [CrossRef]
57. Zhou, J.; Wang, W.N.; He, W.Y.; Zheng, Y.; Wang, L.; Xin, Y.; Liu, Y.; Wang, A.L. Expression of HSP60 and HSP70 in white shrimp, *Litopenaeus vannamei* in response to bacterial challenge. *J. Invertebr. Pathol.* **2010**, *103*, 170–178. [CrossRef]
58. Park, K.; Kwak, I.S. Expression of stress response HSP70 gene in Asian paddle crabs, *Charybdis japonica*, exposure to endocrine disrupting chemicals, bisphenol A (BPA) and 4-nonylphenol (NP). *Ocean Sci. J.* **2013**, *48*, 207–214. [CrossRef]

MDPI
St. Alban-Anlage 66
4052 Basel
Switzerland
Tel. +41 61 683 77 34
Fax +41 61 302 89 18
www.mdpi.com

Applied Sciences Editorial Office
E-mail: applsci@mdpi.com
www.mdpi.com/journal/applsci

Lightning Source UK Ltd.
Milton Keynes UK
UKHW050023020722
405248UK00002B/100